PRIME-TIME AMERICA

PRIME-TIME AMERICA

Life On and Behind the Television Screen

ROBERT SKLAR

New York Oxford
OXFORD UNIVERSITY PRESS
1980

Library of Congress Cataloging in Publication Data

Sklar, Robert.
Prime-time America.

1. Television broadcasting—United States—Addresses, essays, lectures. 2. Television programs—United States—Reviews. 3. Television broadcasting—Social aspects—United States—Addresses, essays, lectures. 4. Television personalities—United States—Addresses, essays, lectures. I. Title.
PN1992.3.U5S56 791.45'0973 80-11760
ISBN 0-19-502765-5

Printed in the United States of America

Again to Leonard and Susan

Contents

PART ONE
Culture and Values on Television

PART TWO
The People and Practice of Television

PART THREE
Television Criticism

Preface

This is a book about television, and also a book about writing about television. In 1976, embarking on a study of television and American culture, I was invited by the editors of *American Film* to report my interim observations in the form of essays and articles for that magazine. Later, I became a reviewer of current television programs for the *Chronicle Review* (later re-named *Books & Arts*). These assignments gave me the opportunity to write regularly about contemporary television; they also forced me to consider how to write effectively about television, in a cultural environment where much is written about television, but little is revealed.

As the most pervasive medium of communications in contemporary American life, television may also be the least understood. The nature of the medium, the machinery of publicity, the structure of the press, all discourage looking clearly at television as a form of cultural communication. The volume of programming is too great, and programs pass too rapidly into oblivion. Industrial competition among the three commercial networks has taken on the character of a sporting contest; given the current tendency to interpret cultural phenomena primarily as spectacle, the machinations of superstar executives in the struggle for ratings have become the most noteworthy aspect of the medium in the eyes of the press.

In the course of writing about television I came to the view that the medium may better be understood as a cultural process—a process by which cultural products are created and transmitted to viewers, who use them, of course, in ways not necessarily intended or predicted. The process begins in the community of creative workers (and their programming superiors) and partakes of their values, working situation, life circumstances. It includes the context of creative forms, traditions, genres, and formulas within which popular entertainment has been created throughout the history of television and its antecedents in movies, radio, variety, vaudeville, burlesque, the circus, pulp fiction, and other popular media—the framework for repetition and variation in television programming. And it can also contain originality and innovation, the contribution of skills beyond the ordinary, the dimension of art.

Production, tradition, aesthetics: these are the elements of the cultural process by which television communicates its meanings, and which television criticism can try to elucidate. These essays attempt in various ways to illuminate the cultural processes of American television—the culture and values depicted on the screen, the people and practices behind the screen—during the latter half of the 1970s.

Foremost I have to thank the editors with whom I have worked during the writing of most of the essays and reviews in this book, Tony Chemasi of *American Film* and Bill Sievert of *Books & Arts*. Each in his own way has provided a climate of confidence and challenge which supports a writer while encouraging him to do more, and better, next time. Hollis Alpert, editor of *American Film,* and Stephen Zito, former editor of *Emmy* magazine, also gave me opportunities to develop my writing on television.

I am grateful to the Rockefeller Foundation for a Humanities Fellowship which, inadvertently, served as the aegis for the inception of these essays, and to the John and Mary R. Markle Foundation for aid for my further study of television; I thank especially Jean Firstenberg, formerly of the Markle Foundation, for her support. John Raeburn of the University of Iowa and Sam Girgus of the University of New Mexico made possible occasions for stimulating discussion of the cultural processes of television with American Studies students. Sheldon Meyer and Leona Capeless of Oxford University Press brought enthusiasm and a critical eye to the process of producing this book. My children, Leonard Sklar and Susan Sklar, have been frequent companions before the television set and unfailingly discriminating viewers. These and many other persons have made helpful contributions to this work.

New York R.S.
February 1980

PRIME-TIME AMERICA

PART ONE

Culture and Values on Television

1. Prime-Time America

They went looking for America, said the ads about those two motorcyclists in *Easy Rider,* but they couldn't find it anywhere. Well, I've been straddling the living room sofa many an evening of late, searching for the same object on prime-time network television, and I've had better luck, of sorts. I found America, a recognizable America, all right, perhaps an all-too-true America, but one you don't often see any more—except, as it turns out, on the most popular and pervasive medium we have. It's the America you used to see in the old tabloid yellow press and in the poverty-row exploitation movies and the carnival sideshows at county fairgrounds, an America of freakishness and sensations treated with solemn import, an America of prurience and violence nicely coated with sanctimony. Why, it was hard to believe I'd found all those old-fashioned American themes on the modern electronic wonder—baby-selling, wife-swapping, juvenile crime, drug-taking, and ripping-off "dating service" customers, old-fashioned themes like that. But on second thought, it seemed obvious that if they were preserved anywhere, it would be on prime-time network television.

You probably knew all this already. Just about *everybody* must know it. The A. C. Nielsen Company tells us that around 60 percent of America's television sets are tuned to network prime-time programs every evening, and since nearly every American home has a television, it means that for a majority of Americans I'm reporting old news. Well, it was fresh news to me, because I've been spending my evenings at the movies these past few years, and there must be some others of you out there, too snobbish or too cheap to own a television set, book freaks or conversationalists or swing-shift workers or early-to-bedders, who will find it surprising, too.

Of course, even the majority can still learn a thing or two; sometimes the things you think you know can fool you. I, for example, thought I knew a little about the people who made prime-time television, having hung around in Hollywood and on Sixth Avenue in New York, and admired the well-behaved, well-groomed men and women of the networks, coolly casual in the West, icily elegant in the East. It never occurred to me that beneath their polished manners lurked the guile of a carnival barker and the brass of a hard-bitten night city editor I once knew, who tried to put some slice of pathos on the front page every morning—a terminally ill child, a fatal tenement fire, a cruelly abused animal—to make his readers' daily burdens seem less severe. It never occurred to me that the Chicago police pressroom from Hecht and MacArthur's *Front Page* had transmigrated lock, stock, and rolltop desk to the beaches of Malibu.

You only have to glance at your local newspaper to grasp my point. Everyone's trying to look like the *New York Times* these days; everyone's become so respectable. The *New York Daily News* is a shadow of its former self, a downright serious newspaper on urban and national affairs. Where are the scandal and gore of yesteryear? You won't find them in the weekly *Star* or the *National Enquirer* any more. Everybody's acquired *taste.** Of course, there are X-rated movies for sex and the others for gratuitous violence, but they're so blatant and raw and disorderly and unredeeming. What you want is a moderate amount of mayhem and murder, a soupçon of titillation, and a square-jawed stance of moral uplift. It's as American as baseball, hot dogs, apple pie, and Chevrolet, and you can only get it these days on television.

* This was written before Rupert Murdoch acquired the *New York Post* and restored tastelessness to the daily press in the nation's metropolis.

But it may not be with us that much longer! There's talk that the sit-com and the action-adventure series are heading for the pop-culture graveyard. What's ahead, according to some media prognosticators, are more prime-time specials, novelizations, mini-series, and made-for-television movies, which can only mean higher tone, more respectability, more taste. Where will we go then for morally sanitized tastelessness?

So come with me while the getting's still good, and take a tour of 8 to 11 p.m. America. I offer you no public television, no specials, no movies, no news, no documentaries, no sports—just a few samples of the weekly fictions that have gilded our world during the 1976–77 TV season. Regarded separately, they lack the essential element network television calls "flow," the unceasing movement of words and images from program to commercials to promotional trailers that is meant to hold the viewer to the set and particularly to one channel, the fast-paced kaleidoscope of sensations that leaves the viewer (this viewer) mind-benumbed after a three-hour ride through Prime-Time America.

Let's begin with the notorious wife-swapping episode on "Maude." This one made the headlines before it hit the screen. The CBS department of program practices rejected the show when it was first produced for the 1974–75 season, but it went on late in 1976 substantially the same, according to its producer, except for cast changes. The controversial subject is broached early in the half hour. A psychiatrist and his wife are visiting Maude and Walter. Looking around the living room, the psychiatrist says, "Your sense of color is marvelous, your drapes are perfect, and I'd like to go to bed with you."

"Maude" is taped before a live studio audience, and, when these lines were uttered, the audience burst into laughter and applause. Laughter, surely nervous laughter, OK. *Applause?* What on earth could be on our collective minds? My sanest guess is that there's a deep element of self-congratulation here. We were present at a historic breakthrough; we had the candor and sophistication to handle the shock; everybody, cast, crew, *and* audience, deserves a cheer. Or maybe everyone's so anxious, mere laughter just won't do.

No, the psychiatrist and Maude do not go to bed together. She's interested, but his wife and her husband are stumbling blocks. The other woman is a terrible drag, constantly sniffling, sneezing, putting drops in her nose—a drip in more ways than one (I forget if they used that line on "Maude," but it's typical of the show's verbal humor). In the end, the psychiatrist confesses that their insouciant sexual adventuring is really a mask, a desperate attempt to shore up a crumbling, unhappy marriage. This pathos draws no applause, but it's what we're there for—you see, swingers really are more miserable. You get your titillation, you get your moral superiority, a half hour well spent.

"Maude" is one of no fewer than eight shows, all comedies, produced for prime-time network television by Norman Lear's companies (not counting "Mary Hartman, Mary Hartman," which is non-network). And though all of them are produced, written, and directed by different people, it's uncanny how very much alike they are. The Lear formula is like the Grand Canyon or the Mount Rushmore of contemporary television, one of the major landmarks of Prime-Time America.

So it shouldn't be surprising how old-fashioned Lear's programs are fundamentally. They could as well be radio comedies—try listening to one sometime without watching the picture. The humor is almost entirely in language and voice inflection—otherwise the actors are mainly mugging for the camera and the studio audience. Nearly all the comedies are built around a gross caricature or stereotype, a kind of holy fool, a lovable sinner, an errant egotist—Maude, Archie Bunker, Schneider on "One Day at a Time," George Jefferson, J.J. on "Good Times"—who gives voice to forbidden, socially ostracized thoughts in a way that excites us without implicating us. As with the "Maude" wife-swapping episode, feelings of titillation and moral superiority go hand in hand.

Lear pulls the old, genteel comedy trick of letting you have your cake and eat it, too. There's a lot of anger and aggression in Lear's comedies, a lot of shock humor on racial, ethnic, and sexual themes—all against the grain of America's pieties and taboos—but the moral equation is always carefully balanced. The more aggression, the more weakness, error, foolishness. Often on a Lear show nearly everybody comes out a little wounded—forgive us, viewer, we've thrilled you with our sex and ethnic talk, but we're just silly dummies, so you can have it both ways.

Within his comic framework, Lear also provides a steady diet of moral preachment—little lessons for the viewer on the right way to live. An episode of "One Day at a Time": Schneider, the sexist superintendent in a building of "menless women," is visited by his nephew, who turns out to be a kleptomaniac. Schneider refuses to see the problem, because he's trying to protect his macho self-image. Ann, the divorcée with two daughters, tells Schneider that the nephew, by his thievery, is screaming for help and needs professional care.

"Send a Schneider to a psychiatrist? I wouldn't even touch their plumbing!" (Laughter and applause from the studio audience. What is it with these studio audiences? Do they go to Lear shows to display their naked anxieties?)

"Would you prefer a prison?"

The Lear formula in action: Poke fun at psychiatry, give us an opportunity to release our fears and prejudices, then the upbeat lesson: Psychiatry is important, it's OK.

Lear deserves credit as a pioneer who brought blacks into Prime-Time America through his comic conventions. "The Jeffersons" spun off from "All in the Family," and CBS, in fall 1976, reshuffled the program from Saturday to Wednesday night to run back to back with another Lear black family on "Good Times"—a significant juxtaposition; first the poor blacks in a Chicago housing project, then, to the tune of "We're movin' on up to the East Side/ We finally got a piece of the pie," the nouveaux riches blacks in a New York luxury high rise.

"Good Times" is a program about lower-class blacks with middle-class values. On a recent episode, J.J. wins $2,500 in the state lottery. After the kids jabber about all the things they can buy, Mama says, "We sound like a bunch of pigs at a trough. That money belongs in the bank." They deposit the money, but Edna, a young girl who lives in the project, appears with a gun and holds the kids hostage while demanding that Mama withdraw the money and give it to her.

"Where do you get off robbing poor people?"

"I'm poor people."

"I got news for you—we're all poor people, only we don't go around robbing our friends."

Mama returns with the money, but the robbery is foiled because of a plot twist involving the illiteracy of Edna's sister and accomplice (the actress, Shirley Hemphill, played a similar character with a similar handicap in another Lear show, "All's Fair," aired just a week earlier—a little too much Lear similarity?). As they hold the would-be robbers for the police, Mama says, "We won't get no police in this neighborhood."

"Yes, we will," says J.J., "tell 'em we white." This was a program where the morality came early and often, and a touch of aggression was needed at the end to achieve the right Lear balance.

Next, the rich blacks. ("Good Times," by the way, carried commercials for perfume, iced tea, a shower massager, a pain reliever, panty hose, breakfast cereal, and coffee; "The Jeffersons" had messages for perfume, luggage, cat food, a massager, a camera, an after-shave. There seem to be some differences in placement of commercials by class content of the programs; rich and poor, however, appear to be equal in the appeal of massagers.)

"The Jeffersons," like "One Day at a Time," "All's Fair," and many other Lear shows, deals heavily in male chauvinism: expressing it, and putting it down. This particular episode featured George Jefferson's effort to torpedo his son Lionel's impending marriage to Jennie, a black girl with a white father. George's prodding gets Lionel to tell Jennie that the man makes all the decisions in a marriage, and Jennie fights back, declaring, "You don't have any rights over this girl." The marriage is off. "I've got my pride," says Lionel. "Pride!" exclaims his mother. "You try sleeping with that the rest of your life."

Finally, his father's caricature of macho behavior makes Lionel see the light. "From now on," says Lionel to Jennie, "we're gonna make all our decisions together."

"Equal partners all the way?"

"I'd rather have my woman than my way."

A little lesson in respect and trust, to go along with the "Good

Times" lessons on thrift and class and neighborhood loyalty. Norman Lear, Professor of Ethics at Prime-Time U.

Lest we get too complacent about Norman Lear's formulas and moralisms, take a look at some other prime-time comedies this past season. "The Tony Randall Show" features Randall as a judge in Philadelphia. In one episode, "Franklin in Love," he proposes marriage to a female judicial colleague named Ellen. She accepts, then changes her mind, because "I don't think I love you enough." She wants companionship from him and probably sex, which she signals by the words "be together" while making sign-language quotation marks with her hands in the air. Randall is willing to "be together"—have sex with her—but he won't guarantee the companionship. A sour, muddled ending presented as a triumph for Randall, as if he won a struggle, despite being rebuffed in marriage.

(A friend with whom I watched the show asked, "Why is her skirt so short?" A judge with an above-the-knee hemline in 1976—a clue to her moral unsuitability?)

"Sirota's Court," a mid-season NBC newcomer, is another comedy about a judge, but the opening shots under the titles don't show city landmarks, as with Randall's Philadelphia; they show an unnamed urban location where crime flourishes on the street and in the halls of justice. This city, says the voice-over, is a jungle. The show has the Norman Lear matter without the manner—there are racist jokes and sexist jokes and contemporary shock jokes about subjects like marijuana (the plot of one episode centered on the arrest and trial of an assistant district attorney for possessing a joint), but they're given to us straight, there's no balance. Racist, sexist jokes are uttered by good guys and are meant to stick. The show is a jungle, too.

Finally, Danny Thomas in "The Practice." Sex jokes, bathroom jokes, thermometer in the rear end jokes, doctors and nurses in the closet jokes. Borscht belt vaudeville for the millions. Another example of the old-fashioned ways of our modern mass medium.

Enough laughing. Let's get violent. Better yet, let's get violent and keep on laughing. One "Starsky and Hutch" episode begins this way: Our plain-clothes pair race to the scene of a laundromat robbery in progress. Starsky strips to his underwear so Hutch can walk in with an armful of clothes, looking like a customer. Hutch blasts away at the holdup men, but one goon grabs an elderly woman as a shield and hostage. Starsky enters with a towel around his waist. The woman screams—whether at a nearly naked Starsky or at her peril isn't clear.

Starsky instructs her: "Bite him."

"I haven't got any teeth."

"Well, step on his foot."

She does, and escapes the goon's grip. Starsky collars him as uniformed cops arrive on the scene. One looks at Starsky with his towel and says, "Who are you, the tooth fairy?"

This bit of comic terrorism leads to a scene of real terrorism. Two more young goons—Los Angeles must be teeming with young, white, Anglo-Saxon hoodlums—lure a retarded girl into an empty bus, where one beats and rapes her. The girl happens to be a protégée of Starsky and Hutch. She tells them, "I hope you beat them up when you catch them. I hope you beat them and beat them."

"It's not that Starsky and I wouldn't like to hit them and hit them," replies Hutch. "But we're police officers, you know. Then we wouldn't be any better than they are."

The episode turns into one of Prime-Time America's endless variations on the *Dirty Harry* theme: Can "the system" provide justice, or must the lawman violate the law to combat evil and gain just retribution? The flaw in the system this time is a politically ambitious young assistant district attorney who doesn't want to prosecute the rapists because he thinks the retarded girl is a weak witness, and he doesn't want to lose the case. Starsky is so angry he grabs the attorney, accusing him of not caring about the girl's rights. "It's 'the system,' " says the police captain, pulling him away. "Most of the time it works."

Starsky and Hutch break the rules and burst into the district attorney's office. They plead the girl's case, and the district attorney is happy to overrule his subordinate—it gives him a chance to quash a potential political rival. The boys haven't had to go outside the law, they've only ignored bureaucratic procedures, and who can blame them for that? They go after the goons, chase them down, and violently punch them out in the process of capture. "What gives you the right to do this?" gasps a battered goon. "It's called justice," says Starsky, "sweet justice. Once in a while it works."

There's a moral to this story, somewhere among the following: (a) "the system" works once in a while; (b) "the system" works most of the time; (c) "the system" works when you step out of line and violate its bureaucratic rules; (d) sweet justice is clouting a rapist in the snout.

Such ambiguity is uncharacteristic of Prime-Time America. Action-adventure shows, no less than comedies, exist to provide you with answers to perplexing human and moral dilemmas. The problems of mental retardation are clarified in "Starsky and Hutch": "What mommy wouldn't love a child who never grows up?" On a "Police Woman" episode, "The Lifeline Agency," about cracking a baby-selling ring, Sergeant Pepper Anderson, played by Angie Dickinson, has this to say about babies who are sold to couples who can't find what they want through legal adoption agencies (presumably, a white child): "Let's face it, the bottom line is that it's no different from slavery."

How to deal with the problem? There's no mention of contraception, for pregnancy seems to be a psychological phenomenon. "She didn't have anyone—that's probably why she got pregnant." Abortion is dealt with only indirectly: An abortion clinic turns out to be a recruiting center for young girls who are encouraged to carry their babies to term and supply them to the ring. The show's explicit answer—presented dramatically rather than didactically—is for unwed mothers to have their babies and presumably to keep them.

The problems of just such a mother happened to occupy a "Family" episode, "The Cradle Will Fall," just a few days later on another network. Selena McGee is a young circus performer who had gotten pregnant in a relationship that later ended. She decided to have the baby and raise it alone. However, when she goes on tour she leaves the child with its father, who has since married someone else and settled down. His new wife, it turns out, can't conceive a child, so the father determines to keep the child. Custody always goes to the mother, says a lawyer, unless moral turpitude is shown. Selena gets caught at the Mexican border with a van full of marijuana—she doesn't smoke, she is conned by "friends." But she's on probation and stands to lose her child.

Selena is a sweet thing but a little woolly upstairs. She manages secretly to take her child from the father's home and is about to run away, but Willie, the son of the "family," who happens to love her, gives her these alternatives: "You could face things like a grown-up. You could run and keep running and pretend some magic would come along and make things right.

"There isn't any magic.

"I don't want to be with someone who won't tell me things [for example, that she got busted, or snatched her kid]. I love you, but I don't think love can grow if people can't talk to each other."

She doesn't run away, but there's enough magic for the father to change his mind and give up his claim to the child.

"Starsky and Hutch," "Police Woman," "Family," and a number of other prime-time shows are set in Southern California. That shouldn't be surprising, since most of the shows are produced there. Television critics occasionally complain that Prime-Time America is exclusively an indoor world, and this certainly holds true for the situation comedies—they're taped on studio sets, and the only sense of place comes from filmed shots under the opening titles—scenes of Chicago or New York or Philadelphia to place the studio sets as part of an actual location. Scenes of actual Los Angeles locations appear frequently, however, in many action-adventure shows. That raises the question: How "actual" is Los Angeles?

Compare, for example, "Kojak" in New York with "Charlie's Angels" in Los Angeles. When you see the World Trade Center as a backdrop in "Kojak" you're seeing something everyone recognizes. Other Manhattan locations may be less familiar, but you have the feeling they are real places, definite streets and corners and stores, and once in a while you know one.

In one "Charlie's Angels" episode, Jill and Kelly were trying to find Sabrina at a certain warehouse on San Fernando Road. There's a warehouse pictured on the screen, but I wonder how many Los Angeles viewers say, "Hey, there's that warehouse on

San·Fernando Road." My son, who was watching with me, in fact said, "Hey, that looks like Sepulveda Boulevard," before he heard it was San Fernando Road. For that matter he could have said La Cienega or Slauson or Long Beach Boulevard or any other light industrial street in the hundreds of square miles of Southern California. Or it might have been in Kansas City or Albuquerque or Memphis—almost any warehouse in America.

The point, I think, is that Los Angeles is a location by default—because that's where the people who make the programs live—rather than for its character. The writers put actual place names in the dialogue because those are the places they know. But Los Angeles rarely comes across on screen as an actual place—those warehouses and homes and storefronts might as well be on some studio backlot.

Who are Charlie's Angels and Sergeant Pepper Anderson and all the other male and female cops and private eyes chasing on the nondescript streets of Los Angeles? With all those studies telling us that prime-time violence makes viewers more aggressive—or, at the least, convinces them America is a dangerous place to live—I can't recall much attention paid to just who are the prime-time bad guys. Are they the deranged hippies and vengeful ghetto dwellers of our contemporary urban nightmares? Starsky and Hutch may specialize in young white goons, but for just about everyone else the answer is, of course not. And it's not just minority sensibilities the producers are protecting. The choice of villains in Prime-Time America is as old-fashioned as so much else in that small-screen world.

For the bad guys are mostly just like you and me—apparently respectable, handsomely dressed, suave of diction and manner, seemingly pillars of the community. But they really aren't like us. Their cool, polite sophistication is just a front. In the "Kojak" episode "Black Thorn," the meat-packing plant headed by dapper Mr. C in his three-piece suit is really a front for dealing heroin—the bags are hidden in beef carcasses in the plant's refrigerator. In the "Charlie's Angels" episode, "Consenting Adults," a high-class dating service turns out to be a front for thieves, and an antique dealer fronts for a diamond smuggling operation—two fronts on one show. In another episode of the intrepid girl detectives, "The

Seance," a psychic fronts for a scheme to rob her patrons. The abortion clinic in "Police Woman" fronts for the baby-selling ring.

This is hardly white-collar crime. It's silk-shirt crime. It fulfills not modern urban paranoia but old-fashioned secularized Puritan paranoia: Nothing is as it seems; the devil lurks in unsuspected places; behind the apparent world of good there's a hidden world of evil. It's not the evil in outsiders that's of any interest—a goon is a goon. It's evil at the heart of what appears to be good.

It takes that old-fashioned cop, Columbo, to see matters in just the right light. In a recent ninety-minute special episode, he's investigating a crime at a family-run private art museum. Two persons have been fatally shot, a burglar and the male head of the wealthy, famous family. Columbo's task is to discover what viewers have already seen, that the burglar was hired by the family's bitter maiden aunt in a plot to murder her brother. She shifts suspicion to her naïve niece by planting a glazed, Renaissance art piece in the girl's room.

Columbo carries the incriminating evidence to the girl's jail cell, where the suspect, ignorant of the art work's value, uses it as an ashtray. That clue leads him to the real criminal. Great fortune did not protect this family from murder, amorality, and ignorance; we knew it all the time.

This "Columbo" episode was called "Old-Fashioned Murder." The old-fashioned vices still thrive in Prime-Time America—along with old-fashioned, sanctimonious moralizing, and I leave to you whether that's a virtue or a vice.

2. The Fonz, Laverne, Shirley, and the Great American Class Struggle

In the spring of my high school senior year, my pals and I pledged to wear white T-shirts until graduation—this was the fabulous 1950s, of course. That plain, white uniform was our emblem of the democratic myth, that we were just common folk, equal and united, when everything around us, test scores, college admissions, our own ambitions most of all, conspired to drive us apart. Very *American Graffiti,* and it was California, too.

For me the moment of truth came when the Lions, or some such service club, chose me a "boy of the year." They asked me to lunch to claim my prize. Would I betray my honor for an honor? Not on your life. Every man and boy wore a suit and tie but me. I kept the faith and accepted my trophy in a short-sleeved, cotton, crew-neck top, white. My pals, I suspect, thought I was nuts.

I remembered that incident for the first time in, uh, several decades while contemplating the Fonz on "Happy Days." In the mythical 1950s' high school world of that ABC television series, Arthur Fonzarelli is the only character who wears a white T-shirt. And it sparkles. It dazzles. It glistens. It gives off the very same aura my friends and I recognized long ago in that symbolic garment—of innocence tempered by experience, of purity tested by reality, above all of an idealized common life. Forget Fonzie's black leather jacket, his ducktail haircut, his tight blue jeans, those fifties' stigmata of a badass hood. That white T-shirt tells you Fonzie is a force for righteousness.

The Fonz is also a working-class figure in a blatantly middle-class setting. Originally, "Happy Days" was about the Cunningham family, a model of comfortable suburban bourgeois living, in a handsome frame house with doors that never seem to lock. Fonzie was simply a supernumerary, a touch of crass to temper the wholesome highjinks of a hardware dealer's offspring. But Fred Silverman, president of ABC entertainment, sensed some new tempo in the public's pulse, so the story goes, and issued new orders to the "Happy Days" producers: Ease up on the wholesome, push crass. The Fonz became the program's hero, "Happy Days" zoomed to the top of the Nielsen ratings and spun off "Laverne & Shirley," a sit-com about two working-class women who toil in a Milwaukee brewery, also set in the 1950s.

"Happy Days" and "Laverne & Shirley," back to back Tuesday nights from eight to nine, are the success story of the 1976–77 prime-time network television season, ranking among the top three shows week after week. Something new seems to be brewing in video land, and it's not simply the working-class hero or heroine. There have been working-class figures on television from Chester A. Riley of "The Life of Riley" through Ralph Kramden of "The Honeymooners" to Archie Bunker, but they and their counterparts are lovable, though trying, buffoons, objects of ridicule whose comic energy is directed as much within as outward. They wear their class like a badge, but class (as opposed, say, to race, sex, or simple intelligence) is rarely made an issue.

The Fonz, Laverne, and Shirley are different. They have their self-mockeries,' but these are leavening features, not the point. They are aware of class and of how it functions in their lives. And they can summon values which, though not reserved exclusively to their own class, seem securely rooted in a sense of class experience. On the first Tuesday of the new year, "Happy Days" and Laverne & Shirley," one after the other, provided striking examples:

A beautiful, smartly dressed young woman, Adrienna Prescott, comes to pick up a car Fonzie, a mechanic, has repaired. He fascinates her, naturally, and she invites him to play at her tennis club.

Fonzie doesn't know the game—until recently tennis egregiously flaunted its exclusiveness—but he survives the match, and romance blossoms. Howard Cunningham, the hardware dealer, gives Fonzie some hardheaded, class-conscious advice: "The two of you come from such completely different worlds. Look . . . you can get along in any set. But can she get along in your world?"

Fonzie takes Adrienna to the high school prom. One of the boys whispers that she's a married woman, and Richard Cunningham reports it to the Fonz. He confronts her. Well, yes, she admits, she is married, but she and her husband "have an understanding. He doesn't tell, and I don't tell."

"I got some rules I live by, y' understand," replies the Fonz, "and one is I don't take what ain't mine, understand?"

There turns out to be a considerable amount of starch in that white T-shirt. Adrienna is invited to split the scene. We all know Fonz is hardly a prude when it comes to women, but he demonstrates he's a man of principle. His action gives some depth to Howard Cunningham's words: The message is that Arthur Fonzarelli's working-class world has a firmer grasp on the moral verities—in this case the Seventh Commandment—than can be found amid Adrienna's affluent, amoral chic.

You could say that this example of class difference is artificial, an overlay, part of the plot but not a fundamental class distinction—obviously, there are moralists among the rich and amoralist grease monkeys. That's not the case with "Laverne & Shirley," where the class conflict is real, because the issue is money.

Laverne and Shirley leave their basement apartment to go looking at clothes in a stylish boutique. "Shirl," whispers Laverne, who is awed by the snooty atmosphere, "We *do not* belong here." Shirley has more front. "We'll see the new styles, colors," she says, "then we'll go down to Woolworth's and buy the same thing." But it's hard to keep her cool when she sees what a dress costs: "Look at the price! It's a year's rent!"

The shop's officious manager bustles over to discourage the déclassé intruders. "We cater to the well-to-do, the crème de la crème," he says haughtily. Laverne and Shirley don't shrink at his aggressiveness, they give it back with both barrels. "Us two girls wouldn't buy dresses here if we were rich and naked," says Laverne. And Shirley makes it personal: "We two girls wouldn't want to buy something from a man who smells like the inside of my grandma's purse."

They stalk out, victors in the verbal battle, but the shop's security man stops them and triumphantly extracts a handkerchief from Laverne's purse. She's arrested for shoplifting and hustled off to jail. (Shirley had noticed egg on Laverne's teeth. In her double anxiety at being out of place in the shop and uncouth besides, Laverne had grabbed a sales item to clean her teeth and then absentmindedly had stuffed it in her bag.) The price she pays for crossing class barriers may be a heavy one.

It's no fun for Laverne in a cell with four seasoned female criminals—"These people do not understand cute and warm," she tells Shirley, "they understand hit, they understand smack"—but, after all, this is a sit-com, and comedy prevails. Shirley summons her strength and invites the shop manager to dinner. The hint is that she promises him sexual favors in return for dropping the charge against Laverne. Ultimately, the plot is resolved by turning the shop manager into a complete fool: He's effeminate but he's a lecher, he's vain, a bully, a snob, unstable. He goes overboard in his class disdain for the working girls—"You can't be nice to people like you," he screams, "the only thing you understand is threats"—and finally, as with the Fonz and Adrienna, the viewer is left with the one simple message: Working-class people are more decent human beings than the well-to-do.

I certainly don't mean to imply that Fonzie and the brewery girls are vanguard fighters in the great American class struggle. But something strange seems to be going on here. Working people, especially working women, are popping up on prime-time television like mushrooms on the forest floor.

CBS offers us the Mary Tyler Moore trio in three separate shows. Rhoda the designer, Phyllis the secretary, and Mary the news editor; as well as Alice the waitress, a series based on the motion picture *Alice Doesn't Live Here Anymore;* and two Norman Lear entries, Ann Romano, the divorcée on "One Day at a Time," another secretary, and Charlie the girl photographer on

"All's Fair." ABC put in during mid-season a replacement series, "What's Happening," a black family sit-com with Mabel Thomas as the mama who works as a housemaid in white households, to go along with "Laverne & Shirley"—not to speak of the humans and humanoids of "Wonder Woman," "Bionic Woman," and "Charlie's Angels." Only NBC seems to lag on the working-class scene, offering only Friday night's standbys, "Sanford and Son" and "Chico and the Man," besides the usual crop of law enforcers.*

The frequent appearance on prime-time network television of people who work, indeed of people in the workplace, contrasts sharply with the good old days, when we used to argue fiercely about precisely what Ozzie Nelson of "The Adventures of Ozzie & Harriet" did for a living—he seemed to be hanging around his suburban dream house all the time. Of course, it may just be a fluke, one of those trends that happen in mass entertainment because producers simply copy whatever another producer has success with. Sometimes it's a mistake to think what we get on television is there for a reason. But let's assume there's logic behind the new image of working people in prime time, and try to deduce what the rationale might be.

Let's begin with the well-known dictum that the purpose of network television is to sell audiences to advertisers. The larger an audience a network can deliver—or, in these more sophisticated times, the larger an audience of defined demographic characteristics, such as age and income—the more it may charge for commercial time, the higher its profits. Ergo, the networks want to program shows that will attract audiences and not drive them away to bed, book, or bottle. Networks have no compunction these days about taking unpopular programs off the air within weeks of their debut. Working people as series' subjects have got to pull audiences or they wouldn't be there. Work, like those old favorites, sex and violence, is suddenly turning viewers on.

We may be witnessing a significant shift in American popular

* As things change, they stay the same. There seemed to be a decline in working class subjects on prime-time television in the late 1970s, when several of these programs left the air. But in 1980 a number of new working class series had trial runs on the networks.

taste. Since the days when Andy Hardy made Louis B. Mayer cry and earned MGM millions, the comfortable suburban setting, white picket fence, broad lawns, sturdy frame house, crackling fire and family by the hearth, has been one of the most powerful dream images in our popular media. Commercial television came along in the late 1940s just as the returning servicemen were buying up raw suburban plots and trying to make their personal Hardy family dreams come true. Through their first three decades, the networks have pandered to those dreams. But sometimes even reality can break in upon life on the small screen.

The 1970s brought us inflation and recession, the worst economic slowdown since the Great Depression. That white frame house in the suburbs is fading from the grasp of those who haven't got it already, and those who've got it are having harder times paying to heat it. Nobody really knows how such important changes in social and economic life affect tastes in entertainment. Personally, I'm dubious that a diet of "escape" is what people want when life grows difficult, and the striking evidence of a decline in television viewing during the 1975–76 season would seem to support this view. Television programs certainly don't "reflect" American society in any precise sense, but to be popular they do need to express, in their various conventional stylized ways, some of the real feelings and concerns of their audience.

And some of those real feelings these days have to do with getting and keeping a job, putting bread on the table, having money in the pocket. The romantic suburban myth is by no means moribund but the mood has shifted. Paddy Chayefsky caught some of the new mood in *Network* when Howard Beale spurs his viewers to shout, "I'm mad as hell, and I'm not going to take it any more!" (And Chayefsky also sensed how much of that untapped anger is directed *against* television.) Maybe it was a similar intuition that led Silverman at ABC, in shifting attention from the Cunningham family to the Fonz, to push for more "hostility" humor on "Happy Days."

What does it mean for viewers when the sit-coms suddenly turn hostile, when they show conflict between the classes instead of sweet accord? Probably no more than a pleasant catharsis, a vicarious thrill to see the rich and the stuck-up get their come-uppance. But if television reinforces attitudes and behavior, as social scientists claim, then to see a television character engage in a struggle and win it may well encourage viewers to persevere in their own battles against inequities.

Take a recent episode of "What's Happening." The show opens with two members of the black Thomas family getting fired from their jobs. Son Roger is canned by a fast-food establishment because he packed barbecue chicken in the same box with vanilla ice cream—funny. Mama Mabel is let go from her maid's job by her white employer on an accusation of stealing a diamond ring—possibly tragic. "We really need those three days' work," she says. It's half her employment; she works six days a week.

Of course, Mabel is innocent. "There are things you should know about, Dee," she says to her daughter. "If something is missing, the maid did it." She goes to an employment agency but can't get another job without a reference from her previous employer. In these straits, Roger gathers two friends and goes to see Mrs. Turner, the employer. It was her husband who insisted Mabel be fired, she says; she can't say any more because she's late for her yoga class.

The boys head for the construction company Mr. Turner owns. Peering into his office, they see a card game in progress and Mr. Turner about to gamble away the diamond ring he accused Mabel of stealing! They burst in, and there's a black man among the card players who takes the boys' side. The black tells Turner that he's guilty of "defamation of character, lack of trust, lack of respect" toward Mabel.

Mabel wants an apology. Turner offers her her job back. She refuses. "I'd rather go hungry than work for you for twenty-five dollars a day." We see her holding the telephone, listening to words we can't hear. Then she says, "Thirty dollars is something else. I'll be there in the morning." Mabel demonstrates her moral superiority, then the primacy of cash values. It seems important that an expression of moral strength on sit-coms not involve personal sacrifice. Fonzie rejects the beautiful Adrienna on moral grounds, but at the snap of a finger he has more high school coeds than he can handle. Mabel is rightly incensed at Mr. Turner's quite rotten behavior, but a five-dollar raise quickly heals her wounds.

This is not to say that being a worker on prime-time network television is as comfortable as being a member of the middle class. In television's new realism, Alice the waitress sleeps on a sofa bed in the living room of her Phoenix, Arizona, apartment. When sister Brenda's boyfriend drops his accordion on Rhoda's foot, her medical treatment at a hospital emergency room is hindered because she doesn't have any medical insurance, and isn't carrying the twenty dollars in cash needed to pay (in advance) for her X-rays.

One episode of "Laverne & Shirley" went even further in exploring the meaning of being a working woman, and of not being a working woman. The girls wake up with the sun streaming through their basement windows. "What good's a beautiful day?" Laverne laments. "We're not going to see much of it in the brewery."

Shirley utters the worker's heresy: "Let's not go to work. . . . Why can't we do what we want?"

And Laverne gives us an updated version of Andy Hardy: "You go to work, you get paid. You don't go to work, you don't get paid. That's the American Dream."

With that hardheaded realism off her chest, Laverne calls in sick. Shirley does the same, and they're off to "do what we want" on a beautiful day.

What do they do? They go to a bakery and buy day-old cookies. They go see *Bwana Devil* in 3D—it's the 1950s, remember. They end up at a playground, fighting a little girl for access to the equipment. It seems the alternative to work is regression, to become children again.

No, something else seems to be happening. Two well-dressed young men are observing them. After much hesitation, the girls get up their nerve and meet the men, eventually inviting them back to their apartment. The possibility of romance vindicates their escape from routine. "You know why we never met these gentlemanly type guys before?" Laverne whispers to Shirley. "It's because we're at work all day."

The gentlemen turn out to be vice-squad officers, dressed up so as to entrap prostitutes—which, of course, they take Laverne and Shirley for. "You're not students, you're not housewives, you're certainly not models."

The show ends happily, of course, but what lingers is the pathos of it. Here are two girls who put caps on beer bottles all day. Breaking out of their daily pursuit of a much scaled-down American Dream—you work, you get paid; you don't, you don't—they find themselves in no-woman's land. In their freedom, they fit no known social category except streetwalker. Skipping work turns the working woman into a criminal. Powerful stuff from the sitcom trade.

It looks like sponsors have a sense of what's going on. On the programs I've mentioned, they've scaled down their version of the American Dream, too. There were a couple of commercials for automobiles (compact cars to be sure) and one for an automatic cooking range (pitched to the working mother), but the rest were for products in the couple of bucks or less category—shampoo, candy, mouth wash, soap, deodorants, toothpaste, razor blades, cough medicine, dog food, cleansing agents. Burger King, it's true, gives you a full dose of the Andy Hardy image in about the first ten seconds of its commercial. A crucial element, however, is missing. Mom doesn't invite her family into the kitchen for a delicious home-cooked meal, she suggests they hop into the station wagon and drive down to Burger King. She probably just got home from work and is too tired to cook.

Diminished dreams and the just plain struggle of America's television viewers to survive seem to have called forth a more "hostile" brand of humor from the networks' prime-time strategists. But programs like the ones I've described, which locate their hostility clearly in the framework of social or economic class differences, are more the exception than the rule. The networks are not in business to sharpen class antagonisms in American society. Though who knows? The episodes of "Happy Days" with Fonzie and the rich Adrienna and of "Laverne & Shirley" with Laverne arrested for shoplifting were the top-rated shows on prime time the week they were screened, according to the Nielsen ratings, and drew audience shares of nearly 50 percent for their time slots.

Network may have it right: Should class warfare give promise of drawing a forty share, we would likely be deluged with social revolution sit-coms.

In the meantime, hostility humor tends to be more diffuse, more general; in a word, safer. There are all sorts of ways to give audiences a charge of anger or resentment without providing too much opportunity to reflect on specific grievances. Intellectuals, for example, make excellent targets. They're snooty and superior, looking down their noses at us common folk, but it's hard to connect them to the cost of living, wages, and working conditions.

"The Mary Tyler Moore Show" gave us a rich opportunity to dislike intellectuals on a recent program, and the added pleasure of resenting Easterners for their snobbery toward middle America. A handsome, mustachioed young Harvard man, *Professor* Carl Heller (accent on the title), arrives in Minneapolis to teach at the university, and impresses the WJM-TV station manager with his pompous pronouncements on books, theater, and movies. Obviously, he's got taste enough to overflow a grain elevator, and he's hired to appear on the local news as "cultural watchdog for the metropolis."

The professor graces his premiere on television with a sweeping attack on Minneapolis as an "intellectually famished, arid, sterile city." Phones in the newsroom start ringing, and Lou Grant blows his stack. Mary tells the professsor, "On the local news we're supposed to appeal to the public, not just the intellectual elite." That just proves to him what's wrong with the news. On his next appearance, he pans the very show he's on for "dull writing, inept staging, high school production methods."

Now Mary blows *her* stack. "What news show did you ever produce, or anything else for that matter?" she says contemptuously. His critiques are simply "sadistic bullying by an arrogant snob." The studio audience expresses its agreement by bursting into applause. The professor gets his just deserts in the classic mode for pompous, self-important people, a cream pie in the face. More heart-felt audience applause.

Hostility of that sort is a dime a dozen on prime-time network television. When you come to think of it, in fact, prime time is just suffused with hostility. Action-adventure shows, of course, convey hostility from start to finish: hostility of crooks against the law and the law against crooks; hostility of cops against district attorneys, politicians, and the system's restraints; the hostility we feel against the muggers, dope dealers, child molesters, church robbers, blackmailers, kidnappers, gun wielders, and other troublemakers who parade across our screen.

Now add to that an increased diet of situation comedy hostility, and you may end up spending entire evenings discharging bile before the tube. It's an exhausting prospect. There's something about watching television that seems to deplete rather than to invigorate the viewer. At a movie like *Rocky,* you cheer the gutsy underdog and leave the theater charged up, walking on air. Maybe live studio audiences get that feeling; they seem to applaud all the time. Maybe it has something to do with being part of a collective experience.

Maybe it also has something to do with the nature of television comedy. There's a difference between comedy and situation comedy. Prime-time television has oodles of the latter, mighty little of the former. It's the business of situation comedy to keep its humor within bounds. There's an orderliness, a moderation, to the sitcom formulas—the humor doesn't make you wince, cry out in pain, guffaw, fall on the floor in helpless hysterics. It doesn't reach the far ends of the comic spectrum; it's the kind of humor that studio audiences are as willing to applaud as to laugh at.

Comedy is out to break down boundaries—to astonish you, embarrass you, gross you out, give you a fresh vision of familiar commonplaces. Compared with comedy, sit-coms are highly cerebral; comedy hits you in the gut.

"The Carol Burnett Show" is the prime example of comedy on network television these days. More often than not Burnett snares the rich and pretentious in her comic net. But when she casts her eye on the working class she can make Fonzie, even Laverne and Shirley seem sentimental dolls.

In a recent skit, Tim Conway and Burnett played a working-class husband and wife. The theme is the "Total Woman" concept, with the husband demanding that his wife greet him in "Total Woman" fashion when he comes home from work. Home is a veritable sty, with dirty handprints on the refrigerator and kitchen cabinets, and tears in the sofa upholstery, with the wife a slattern out of a George Price cartoon, wearing bulbous green earrings, and with an anchor tattooed on her arm.

This, of course, is the comedy of exaggeration, but it speaks to the issues of class difference in a more direct way than Fonzie in his white T-shirt or I, back in my high school senior year, thought I was doing in my white T-shirt. When the working husband on "The Carol Burnett Show" arrives home after work you can be sure his T-shirt is no more dazzling and sparkling white than the smudged refrigerator door in his kitchen. The farce comedy of Carol Burnett reminds you that the new sit-com realism about working people is not real enough to include the dirt and grime of working life.

3. The Backlash Factor: Reflections on Television Violence

It's ten o'clock, and I'm circumnavigating the channels. "Charlie's Angels," an adventure series, on ABC. "The Blue Knight" on CBS. "The Quest," a Western, on NBC. A split-second evaluation of images and I settle on "The Blue Knight"—a glimpse of smog-shrouded L.A. City Hall has plucked a personal chord. There's a slow bit of business with a cop and a panhandler who wipes windshields, and then suddenly it's night. The cutting tempo quickens, jazz on the sound track, a man in an expensive station wagon hails a streetwalker, they gesture and leer, she gets in the car, another man sneaks in the backseat, holds a gun to the driver's head, they drive off, the driver has his own gun, a shot, a crash, screams. It's ten past ten, and crudely, without characterization or subtlety or explanation, a man is dead.

This is a random moment of dramatized violent death on television—one of several dozen every week, a hundred in a month, more than a thousand by the year. It's not typical, there are "worse" and "better" depictions, more or less bloody, cruel, senseless, shocking, but it's symptomatic in the one-dimensional nature of victims and victimizers, their motives, gestures, expressions, and the almost joyful energy of the music and the editing, on a show which is otherwise visually plain and slow.

No more controversial issue bedevils American television than the question of what to do about violent scenes like this. Do they actually stimulate some impressionable young people and adults to commit similar acts of violence? Do they convey an unrealistic impression of the omnipresence of violence in American society, encouraging anxiety and fearful behavior quite at variance with actual social circumstances? Or are they merely a necessary and time-honored convention of storytelling, less gruesome by far than Homer, reaching audiences that in other periods thrilled to B-movies, pulp detective stories, dime novels, or thronged to bear-baitings and public hangings?

These are old questions, subjects already of several thousand research reports and countless editorials and irate letters to network presidents. But during the Bicentennial year—perhaps because of it—they have acquired fresh vigor. In the year of national self-congratulation and patriotic good spirits, violence on television seemed to many a particularly unwarranted affront to American pride and social character. From a leading advertising agency came a startling presentation highlighting the excesses of violence (and sex) in the media. A corporation took a public stand disassociating itself from "programs, publications, movies, or events involving excessive violence, sex, or matters of poor taste." The national Parent-Teacher Association announced a campaign against violence on television. A new wind seemed to be stirring some last straws.

Violence on television: You can abhor it emotionally, deplore it morally, condemn its failures in aesthetic terms and even as story, but there is one blunt truth no one should avoid—television, and other popular media before it, have consistently dealt more openly with the realities of violence in American society than the universities and schools, the churches, politicians, and most of the press. This fact accounts for no small part of the persistent effort to censor and control all major media in their turn, movies once, comic books once, television now. Anything which depicts American society other than as humane, well-ordered, gentle, kindly, moral, benign, and just should not be communicated—opposition to violence on television often follows from a premise as simple as that. Norman Mailer recently suggested that half of America lives in the

19th century and half in the 20th, and the attack on television violence has in considerable part the character of a 19th-century grudge against the 20th for being born.

This makes it inordinately difficult for 20th-century critics of violence on television to find common ground with the moralists whose roots lie in another era. They are forced to take a civil libertarian stand, deploring the message while defending to the death the messenger's right to say it, and wondering all the while whether commercial television producers and powerful networks deserve such ringing declarations of principle. Or they can revert to a more businesslike attitude, conceding that since principle no less than art and story is so difficult to define and implement in the mass media context, it works more effectively to judge such issues, as William James said in his essays on pragmatism, by their "cash value."

The J. Walter Thompson advertising agency has taken the latter tack in its unprecedented multimedia presentation called "The Desensitization of America." This forty-minute montage-collage of images and sounds from popular music, books, magazines, films, television, and urban smut districts was intended principally for in-house use by JWT's staff and clients, but the extraordinary response it evoked among advertising and media professionals, as well as interest generated among the public by press accounts, has led to considerable notoriety and wider dissemination. The notoriety is likely to increase as JWT makes available the results of its national survey of adult TV viewers' attitudes toward violence.

The preliminary findings, reported by Ron Sherman, manager of JWT's New York office, at a staff screening of "The Desensitization of America" (to employees who had seen the presentation as many as three or four times before), are already rippling ominously through the media world. A full 40 percent, two out of every five respondents, said they avoid watching television programs they consider too violent. Twenty percent of the men and one-third of the women said they prevent children from watching such programs. And then comes the pragmatic kicker. One out of every ten respondents has contemplated boycotting products advertised on programs he or she considers excessively violent, and *fully 8 percent said they had actually refused to buy the products advertised on the disapproved of shows.* (When JWT released its results at a national meeting of advertisers, this figure was revised downward considerably.)

A potential 8 percent loss of sales just by placing one's commercials on an offending program! No wonder JWT is concerned, and wants its clients alerted, too. "We are not leading the charge against TV violence," says Sherman. "We are just leading the investigation." But with data like that it's hard to tell where investigation leaves off and condemnation begins. No advertising agency wants to place its clients' products in a programming environment that could actually turn off sales. A few small gestures in the supermarket—taking one brand of canned tuna instead of another which advertises on disapproved of programs—could have a snowballing effect that alters television content more swiftly than any other form of protest.

The public is never wrong, said the movie mogul, Adolph Zukor. Don't get any illusions about your infallibility, but consider that your power, even in this age of mammoth media, may be greater than you think. It consists almost wholly in abstention. You may exhort, explain, persuade, legislate, and get nowhere. Stay away, and your absence will be heard. It is a negative power, but power nonetheless. A few months ago Les Brown was writing in the *New York Times* that television had become a failure-proof business—the networks had to turn away advertisers because all their time slots were sold out months in advance of the new season. A little consumer resistance could turn the most iron-clad media dictum into a cloud of smoke.

No one else dies on "The Blue Knight" this particular evening. Pausing at 10:15 for three commercials, at 10:30 for three more, plus two network promotions, a local news promotion, and a public service ad, at 10:45 for another three commercials, at 10:54 for yet three more, and ending, just before the final logo, with a paid political ad, the program devotes its remaining time to tracking down the victim's killer. The only substantial violence thereafter is verbal—criminal types calling the cops "pigs." When the cop-on-the-beat chases the fleeing murder suspect, he doesn't pull a gun, he outruns the kid and collars him. A bit of fantasy to leaven the Boyle Heights realism.

Dramatic tension comes not so much from the chase as from what might be called caste conflict within the law enforcement world. The cop-on-the-beat argues with detectives and with the district attorney about their willingness to frame the panhandler for murder: "The legal system—let's not go into that," says the DA. The detectives are upstaged by the FBI on a robbery bust that helps finger the killer: "You know the Feds—we do the work, they take the credit." Resentment against higher authority, and a sense of powerlessness to do anything about it, pervade the story.

The program ends on a pair of bitter ironies. The panhandler, saved from a murder rap by the cop-on-the-beat, turns on his benefactor, because a public defender has told him the cop failed to inform him of his rights. And the murderer, who can't be tried on the crime he did commit because of insufficient evidence, will be framed on a federal robbery charge, of which he's innocent, because of a coincidence that provides evidence enough to convict. "The Blue Knight" depicts a chaotic, disordered, irrational system of justice, in which right triumphs barely, and by ambiguous means. If I were a moralist, I'd want to pay as much attention to the tone, the ambience, the ideology, if you will, of programs like "The Blue Knight," as to a few seconds of exciting violence.

Which brings up the question of moral discrimination. How do you decide when violence crosses the line from "just right" to "too much"? What religious, ethical, social, aesthetic, and commercial judgments do you bring to bear? Seventy-five years of complaints against movies have demonstrated no consensus among the moralists. Many wanted to eliminate all crime and sex from films; but without crime and sex as subjects, said Martin Quigley, one of the authors of the Motion Picture Production Code, there wouldn't be any popular entertainment. He wanted to preserve those subjects for the movies, but to ensure that criminals and sexual transgressors were made to pay.

So there's also the issue of context. Certain episodes, possibly offensive or provocative in isolation, may exert a positive moral influence when viewed within the mise-en-scène of an entire work. J. Walter Thompson's multimedia presentation ignores this aspect of the debate. It shows, for example, one after another, the three most sexually explicit scenes from the motion picture *Shampoo,* and they appear shocking and palpably unpleasant. Yet one could make a convincing case for the artistic quality of that film, for the value of those scenes in context, and for an interpretation of the film quite different from the implication JWT conveys through its excerpts.

Despite the enormous and as yet unexamined complexity of these issues, the JWT presentation has impelled one of its clients, the Samsonite Corporation of Denver, manufacturer of luggage and outdoor furniture, publicly to draw the line at placing its ads in unacceptable programs. On the forbidden list are "programs, publications, movies, or events involving excessive violence, sex, or matters of poor taste," as well as those media entertainments "whose story line is known to involve sex, violence, shock, regardless of the degree of editing."

Peter R. VanDerNoot, Samsonite's director of public relations, seems to play down the novelty of the corporation's policy statement when I speak with him about it. "That's the kind of company we are," he says. The statement is a "reaffirmation of our way of doing business." Nevertheless, a press release accompanying the document describes it as announcing "new criteria" for placement of its ads, and the formal statement concludes:

"By establishing this policy, it is our hope that the communications and entertainment media will give positive attention to the reduction of violence, and other elements in poor taste, in the development of future media presentations."

I ask VanDerNoot for examples of programs that do not meet his company's standards, and of others that do. He offers "Starsky and Hutch" as a program that deals excessively with violence, "Columbo" as one dealing "with the same type of material but not excessively violently." A corporation thus showed itself willing to take a specific stand on questions of taste and morality. The specific grounds for that specific stand, however, remain unexplained.

Ten o'clock again. Neither "Starsky and Hutch" nor "Columbo" is available for comparative purposes, so I give the new NBC series, "Serpico," a try. It has similar blood lines to "The Blue Knight," sired by a best-selling book out of a feature movie, but it's soon clear the differences between the two shows are as sharp as the contrast between East Los Angeles and Upper East Side New York. "The Blue Knight" is a blue-collar cop; Frank Serpico jogs in Central Park in a Mostly Mozart sweatshirt.

They hold back the first death in "Serpico" until 10:20. By that time the victim has been seen several times, neither he nor his murderers are perfect strangers. His executioners march him down an alley and turn into a narrow stairway between two buildings. The camera suddenly withdraws. You can feel the panning movement almost physically, an averting of the eyes. Blood-curdling screams on the sound track, on the screen a blank brick wall. Then a slow, almost reluctant dolly up to the stairway, a glimpse of a corpse, sprawled head down on the step. Cut to a commercial for pizza.

Here we have a new definition of camera-shy—it's the shy camera. Can this diffidence be a product of the pressures on producers and networks exemplified by the JWT presentation, the Samsonite statement, and more diffused public protest? It doesn't deter "Serpico" from further violence—three more deaths, by my count.

It's a good thing Serpico has been jogging in the park in his Mostly Mozart sweatshirt. He's been set up to be bumped off by the numbers racket. The hitmen are waiting for him on the sixth floor of an office building, but he does a smart thing. He presses the elevator button for six, darts out before the door closes, and dashes up the stairs as swiftly as the elevator rises. No sooner do the hoods discover they're blamming away at an empty elevator than Serpico appears in the stairway behind them, gun in hand, ordering them to freeze. They shoot, he blasts one, the other runs. Rather than shooting at the fleeing figure, the hip cop runs after him, catches up, fights, and knocks him out. Cut to an ad for sanitary napkins.

Almost the same circumstances recur in the program's climax—an ambush, an exchange of shots, two fatalities, one bad guy running, Serpico disdaining to fire, overtaking him, punching him out. Was the real-life prototype for this hero an Olympics-class sprinter? No matter. I only hope these depictions of police behavior in "The Blue Knight" and "Serpico" don't encourage youths to turn and flee when some law officer yells, "Freeze!"

"Serpico" has its roots in Hollywood's old B-thrillers and pre-Peckinpah Westerns: no blood, no gore, plenty of gunplay and fistfights, clean deaths without visible wounds and lots of time for final statements by the dying before their eyes shut and heads fall limp. These conventions bring to mind another of the classic conundrums of the violence debate: Does this old-fashioned, purified depiction of violence affect audiences any differently than more explicit violence?

A brief wire service item in the paper begins: "The Parent-Teacher Association plans a year-long campaign against violence on television that could include national boycotts of products and programs." This leads me to call Carol Kimmel, national president of the PTA, whom I find at home in Rock Island, Illinois, fortuitously, following a week-long seminar in Chicago for the ten members of her organization's special commission on television violence. They had heard academic experts, network representatives, people from advertising agencies (including J. Walter Thompson), an FCC commissioner, leaders of Action for Children's Television and the National Citizens Committee for Broadcasting, two groups leading the campaign against violence on television. Opinions, of course, conflicted.

The PTA is no Johnny-come-lately to the debate over media violence. "We have been concerned with what's happening on the media, oh, since the days of silent pictures," says Kimmel, with a short laugh at the long struggle her words suggest. With that background the PTA might qualify as one of the most persistent voices of 19th-century moralism in the land. When it comes down to specifics, however, such labels seem less userful—they help to clarify the nature of cultural conflicts, but they rarely do justice to the complex aims of the combatants.

The obvious goal of the PTA is to reduce the amount of violence on television; from another angle, that means supporting "good" programs, and the even more difficult task of trying to envision what a desirable daily television diet might be. Another goal is to get the public to take more responsibility for the content of television programs; but the opposite side of that coin is less clear if it means taking away some of the networks' power over programming.

Getting someone, anyone, to take responsibility for media power is like forcing him or her to hold the proverbial hot potato. The networks only want to please the vast majority; the advertisers and their agencies only want to get their message across. They all credit the public with make-or-break power over television content. If not enough viewers like a program, out it goes! But this familiar stance completely ignores the fact that the public can only endorse or reject, if has no access to the process of planning, conception, creation. Curiously, when the networks invoke the myth of the powerful viewer, they invite the disgruntled to take extreme remedies—if you don't like what you see, it's no good complaining, turn off the set. They're betting on the belief that most people would rather watch something than nothing.

Implicit in the PTA's campaign is the conviction that if such an indiscriminate need for television ever existed among the American public, that yearning may be coming under stricter control. The organization cites survey research which indicates that nearly two-thirds of adult respondents hold the view that there's too much violence on television. When confronted with evidence that ratings for violent programs remain high—seeming proof that viewers will swallow their dislike and keep their sets turned on—the PTA suggests a simple answer: pollsters speak to adults while children watch the shows.

It seems clear that the PTA is prepared to encourage its constituents to take the step the networks seem to condone: If you don't like it, don't watch it, and don't let the kids go near it, either. A boycott of products advertised on these offending programs is not ruled out, but it's a much more controversial step, and further down the road. "We are talking about a tremendous industry moneywise . . . ," says Carol Kimmel, leaving the sentence dangling, with the possibilities a little too disturbing to voice. But a boycott of programs may be as effective as a boycott of products, though it may take longer to make itself felt.

Reformers are often too preoccupied with attacking what they don't like, to think about what they would do if they succeeded.

What would you put on the air if the power were yours? Kimmel answers that she wants more variety in television programs, more moderation, "more realistic treatment about the country." When the networks flood the screen with documentaries on African wildlife, you will know the reformers have won.

Moralists rarely want to change the status quo, they simply want to trim its rougher edges. There is an instructive precedent in the movie field. In the early 1930s the Roman Catholic Legion of Decency organized to boycott offensive movies, and were joined by scores of Protestant and Jewish groups. Hollywood capitulated, the Motion Picture Production Code severely restricted movie depiction of sex and violence. The outcome was that the movie industry for a time grew wealthier and more powerful than ever.

The precedent is only a tenuous one. Television has no antagonist as well organized as the Legion, no leader like the movies' Will H. Hays, no coherent code waiting to be enforced, as the movies did, no economic crisis like the Great Depression to weaken its position. But it does have federal regulation, and a great deal of new technological possibilities not yet fully explored. The capacity exists for many more sweeping changes than occurred in the movies; the current deployment of forces suggests there may be much less.

4. A Word from Our Sponsors

At the Midwestern state university where I once taught, a student called one day and alerted me to a special screening. "Get there quick," he said. "It'll be jammed." I rushed down and found one of the last free seats. By the time the lights went down, students packed the aisles and stood three deep at the rear.

When the first images appeared, the audience clapped, whistled, hooted with delight. Many knew the dialogue, and spoke it out loud in unison with the sound track. Thirty seconds later, however, the image faded to black, only to be replaced immediately by a new one. More cheers of recognition and approval. Another half-minute epiphany took place. And another. And another. The ritual went on for the better part of an hour, and enthusiasm began to wane only at the end, when the dialogue shifted to Japanese, Danish, or Hebrew, and the students could no longer speak the lines.

We were watching a compilation film of outstanding television commercials, American and foreign, winners of the annual American TV & Radio Commercials Festival (Clio) Awards. The spectacle would have warmed the cockles of a Madison Avenue heart faster than a two-martini lunch. Of course, those students may have all been advertising majors, exhibiting a knowing preprofessional pleasure in the best products of their chosen trade. But I doubt it. I think they loved those well-made commercials the way we love a well-told story, and especially the way we love a well-told joke.

For the vast majority of these Clio Award winners were funny, if not a little absurd. They poked us playfully in the ribs, saying: We're so confident of our product we can have some fun with it, with you, among ourselves. We can rise above the tacky need to *sell* you something. We're not pushing cars or home insurance at you, we're sharing a style, a winning attitude. The insurance and the cars are just things you buy if you want to show you've got some style. C'mon, baby (wink), let's laugh all the way to the bank. . . .

At some level, I suspect, the students' enjoyment of these superior comic commercials was heightened by a certain feeling of complicity. They sensed that the jokes we laugh at the hardest are the jokes where we ourselves are the goat.

They're not laughing on the way to the bank any more. And I wonder whether this year's crop of Clio Award winners will be greeted by university students as if they were "The Best of Groucho." Times have changed. When my television set these days shows me people walking into banks, they're angry. They can't cash a check. Can't find out their balance. Can't get a loan. Can't get anybody to talk to them. They darn well better find themselves a new bank that'll give some decent service.

No one's laughing over at the supermarket, either. Have you seen what's happening to prices lately? Coffee, for one, has gone through the roof. That nice man who stamps the prices on the coffee cans commiserates. "I have to put these prices on coffee," he says. "But you don't have to pay it." He suggests we try a brand of tea—the coffee drinker's tea.

And here's Frank Borman, the former astronaut, now president of Eastern Airlines, coming on in the middle of "Mary Hartman, Mary Hartman." Has he got my sunshine? No, he's got some dark truths to admit. "In the past," he confesses, "Eastern's reputation for service wasn't the best." But he promises better days ahead.

Times have suddenly gotten tough in television's world of thirty-second comedy dramas, the commercials. Handsome foreign gentlemen still try to sell you expensive cars, and there's still many a laugh to be found when it's time for a word from our sponsors, but money is tighter in our wallets and pocketbooks, and it's gotten tighter all the way up the line.

CBS put on a special, "The People's Choice Awards," which brought us glimpses of every favorite television and movie personality you might ever want to see, from Farrah Fawcett-Majors through Sylvester Stallone, and ended with a comedy routine between Bob Hope and former President Gerald R. Ford. A prestige show if there ever was one, and it produced a forty-three share to lead that night's Nielsen ratings. But during one whole hour the advertised products were Tide, Pampers, Gleem, Scope, Crisco, Prell, Joy, and Sure. Hardly an item above $1.98 in the lot. It looked like the commercial lineup on the afternoon soap opera. (Perhaps the agency time buyers thought the audience demographics would be about the same.)

To the networks and the local stations, of course, commercials are the bottom line of what television is all about. The whole vast commercial television system is founded on advertising, on the willingness of business firms to pay networks and stations to air their commercial messages.

And in the harder times of the mid-1970s, it remains more than ever a seller's market. Slow times in the economy may make it imperative for advertisers to step up their ad campaigns to overcome increased consumer resistance. The number of minutes networks sold to advertisers were up 5 percent in 1976 compared to 1975. Demand for advertising time exceeds the supply of minutes the networks set aside for commercials.

How these minutes get filled, what gets said, and when and where, are subjects of intense planning and scrutiny among sponsors, ad agencies, networks, and stations. By everyone involved with television except the viewers and the critics. As a very round rule of thumb, say there are seven commercial minutes an hour in an average station's twenty-hour programming day—a low estimate. That would mean that the total minutes of television time devoted daily to commercials would almost surely surpass the total minutes of the morning, noon, evening, and late news programs.

Yet news is one of the most intensely debated aspects of television programming, while hardly a thought is given to commercials. Books on television news reporting cover several shelves, *TV Guide* devotes special columns to it, politicians pontificate about it. Most books on television, meanwhile, don't even have the word "commercial" in their index.

That may be because commercials are accepted by viewers as well as by networks and stations as part of the basic rules of the television game. Surveys consistently show that more than two-thirds of viewers believe that commercial interruptions are a fair trade-off for the entertainment they get. A majority not only finds commercials informative, it rates some commercials as more entertaining than the programs they interrupt.

On the other hand, one among those indefatigable research teams paid attention not to what viewers said but what they did. It discovered that during the time sets were turned on, barely half the tested viewers actually looked at commercials. The other half had presumably leaped up to visit the refrigerator or some other common household device.

Pity the poor advertiser and his ad agency. Half the viewers don't watch commercials and the other half, so additional surveys report, have a hard time recalling what they saw. Maybe the viewer isn't the only goat of the joke.

The latest cross advertisers have to bear is the product comparison. In the good old days, all you had to do was line your product up against Brand A or Brand X. Conclusive evidence right there on the screen demonstrated that your product did whatever it was supposed to do—soften or whiten or foam or work fast—better than the anonymous competitors.

But several years ago the Federal Trade Commission began to insist that the actual competitive brands be unmasked in commercials, that names be named. The idea was that this would reduce the confusion caused by grandiose claims based on information withheld—the consumer would have actual comparisons rather than comparison by innuendo, and would be able to make more rational purchasing decisions by brand name.

So comparisons play a ubiquitous role in commercials these days, and it's harder to turn them into jokes. Pepsi challenges Coca-Cola, Viva takes on Bounty, Dynamo battles it out with

Tide, Kool-Aid vies with Hawaiian Punch, Renuzit goes into the ring with Glade, Volkswagen Rabbit pits itself against the entire automotive world. The spectacle would exhaust the lexicon of a hardened sports-page headline writer.

During a commercial break on a 10:30 network show, a "real person" tells us how much Tide means to her life, loves, and laundry. A half hour later, on another channel, another "real person" confesses herself a former Tide user who has found happiness with Dynamo. What hath the FTC wrought? Will Ms. or Mr. American Consumer be enabled to make more rational choices in the supermarket? Consumer research results recently released by an advertising agency (which may of course have its own ax to grind) assert that the females tested were more likely than not to misidentify the sponsor of a competitive commercial, naming the denigrated, nonsponsoring brand. Comparative commercials seem to function as free promotions for the brand the sponsor would have us shun.

Any ad person could have told the FTC (and probably did) that the American consumer has not yet evolved to a state of rationality, and with luck may never do so. From its misty origins in the distant past of capitalism, advertising has based its premises more on the libido than on the intellect, the solar plexus than the cerebrum. Have you vexing anxieties? Unfulfilled yearnings? Men, do the girls snicker when you walk by? Women, do your beaux unaccountably fail to call a second time? Maybe your scalp, or your breath, or your skin, or, perish the thought, your odor, denies you the pleasure you seek. Simple remedies lie at hand. Merely shampoo, gargle, lather, spray. Happiness will be yours.

Vast industries are built on the quicksand of such promises. Let the FTC tamper with temerity.

Inquirers into the state of the American psyche could save themselves a trip to the Iowa cornfields by pondering the state of the American television commercial. For example:

In the midst of the local late-night news, I am suddenly dazzled by the sight of a green upland meadow, a mountain stream glistening with reflected sunlight, four backpackers climbing into my field of vision. They stop, they let down their packs. What's going on? I knit my brow, I am curious. Oh, they are hungry! The man in front takes out a loaf of bread. It is Beefsteak bread; it is as good as beefsteak to a hungry backpacker. He shows me the label, then the sliced loaf inside the package. Cut away to his companions—two are removing their caps, they are letting their golden, shiny hair fall to their shoulders, they are beautiful young women, they are smiling at each other.

They are gone. They are replaced by an angry young man complaining about his bank. He gives way to a lovely young woman cuddling a puppy and talking about the virtues of a dog food. She departs, and a number of airline ticket envelopes seem to be falling like dominoes, leaving only American Airlines standing. Time for sports. Montreal defeated Atlanta, the Harlem Globetrotters. . . .

My mind is still on those two young ladies up on the mountain meadow. I have seen them only for a second or two, perhaps, at the outside, five seconds. I have probably not accurately described the images I saw. I have not run that commercial backward and forward on a Steenbeck, like a proper scholar, stopping every frame. I have only a memory of a one-time viewing, like the average viewer the commercial is trying to reach. Why was I surprised that they were women when they removed their caps? Was I simply inattentive when the commercial began? Or did the reflected sunlight, the backpacks, the caps, the parkas, initially mask their gender?

Commercials sometimes play little tricks with the genders that way. In this case you get the message that Beefsteak bread is a he-man's meal, then the sudden revelation that the two ill-defined companions are actually luscious objects for virile attention. It happens so fast that it leaves plenty of room for the viewer's fantasy to soar.

Some pundits have talked of television as a medium of dreams, and what could be more dreamlike—swift, surprising, illusive, unworldly—than a commercial? How simple it is for the men and women who make commercials to tap our secret feelings. Certain of my appetites have obviously been touched by that commercial. It's not clear, however, whether one of them is the appetite for bread.

I have asked people who make commercials whether they're consciously striving to attain an impact on viewers similar to the impression that the Beefsteak bread commercial made on me. I guaranteed them anonymity, so they could tell me the truth and still go on making commercials. The answer I got—I'm not sure I believe it—is that viewer manipulation is not planned overtly, but rather unconsciously.

People don't sit around conference rooms, so I'm told, and say, "Let's peg this to the male fantasy of being with a woman in the wild." Rather there is a good deal of implicit, unstated understanding of the value of "secondary associations." If you like the "secondary" aspects of the commercial, you're liable to like the primary aspect, the product. The name of the product itself, Beefsteak bread, is clearly designed to elicit secondary associations. The wilderness atmosphere surely is a positive secondary association. As for anything else in the commercial . . . well, it's up to you.

Those comedy commercials that convulsed the Midwestern student audience were all based on the value of "secondary associations"—like the commercial, like its ambience, like the people in it, you'll like the product. Some commercials wouldn't even name the product until the last few seconds, on the premise that a funny little story would make you so receptive a minimal soft sell would do the trick.

Secondary associations still abound in the world of television commercials, but you hardly ever see minimal soft sells any more. Medium soft sells are more like it. Coca-Cola ads for example, with their simple, reiterative lyrics—"Coke adds life"—over images of young people having fun in exciting cities like New Orleans.

But a soft economy has brought back what might be called a medium hard sell (on the assumption that maximum hard sell is a description reserved for live commercials from used-car lots during the late show). The medium hard sell eschews the fancy camera movement, rapid cutting, and dazzling sound tracks of the Busby Berkeley school of filming commercials—Sugar Free Dr. Pepper commercials may be among the sole survivors of that particular breed. Its eye is on the product. There may be only three or four shots, the product is front and center, sound track emphasizes the product slogan, simple and repeated.

Allusiveness is out. Many commercials have ceased to be comedy-dramas and have turned into visual essays. Tell 'em what you're going to tell them. Tell it to them. Tell them what you just told them.

"Aim *fights* cavities. *Aim* fights cavities. Aim fights *cavities.*"

That doesn't mean, of course, that they can't still contain plenty of "secondary associations," can't still speak to the psyche. The message, however, may be somewhat more blunt and obviously relevant than the putative linkup between sexual fantasy and Beefsteak bread.

These are the commercials that aim at the avoidance of pain rather than the achievement of pleasure; or better, the achievement of pleasure through the avoidance of pain. Take, on a relatively painless level of pain, the Cascade commercial. The young housewife is suffering the pain of embarrassment because her dishwasher soap doesn't get her wine goblets clean. "They were spotty," she tells her husband. "What a nuisance." Not to speak of the social disgrace they could cause at one's next dinner party.

Mom saves the day. She arrives to tell her daughter about Cascade. Sure enough, Cascade gets the goblets "virtually spotless."

These poignant domestic scenes are enacted in the midst of "Another World," an NBC afternoon soap opera, and indeed another world from the two men and two women on a mountain during the local late-night news. Cascade and its ad agency don't expect people like me to be exercising our fantasy lives at that particular hour. They have two particular audiences in mind, and all honors to the writers, directors, cameramen, and editors who can hit two birds with one well-aimed stone.

They've snared the young homemaker aspiring to an affluent life-style and social success—aspiring also to a husband who gives a damn whether the wine goblets are spotty or not. And they've captured every mother of a daughter, not to speak of every daughter of a mother. How effortlessly mother advises, how gratefully daughter receives!

Ah, but try the relationship the other way around. What happens when daughter is the giver, mother the recipient? Try baking your mother a piece of cake. Better make it out of Moist and Easy Snack Cake Mix by Duncan Hines. Then, of course, mom's lavish with her praise. Dad loves it, too . . . and the whole doggone extended family. It's remarkable how much family psychodrama can be packed into a few dozen frames of film.

The mother-daughter encounter is perhaps the most emotionally charged dialogue in a whole subgenre of television commercials, the initiation of the neophyte. These are the commercials in which information and persuasion are part of an internal rather than an external exchange—not addressed directly at the viewer, by actor or announcer, but shared within the commercial by two or more performers.

Wisk, with its crusade against "ring around the collar," offers some of the brasher commercials of the type. It's got to convince viewers there's a problem before they may turn to the proferred solution. Pampers can afford to be subtler. When you're pregnant, you don't have to be persuaded of the need for diapers. But for Pampers the sell has to be toned down, because you can't address viewers as if pregnancy were a situation they all shared, like dirt or hunger.

Pampers does it by showing pregnant women talking to one another. For one it's her first pregnancy, the other has been through it all before. That's why she recommends Pampers. "Boy, I'm really glad I talked to you," says the first-timer. "Ask any mother about Pampers," says the voice-over. There's confidence: We don't have to sell our product, we'll let others do it for us.

On an afternoon soap opera, the Pampers exchange took place between two white women. On a prime-time network show one woman was black, the other white, and the commercial ended with the white woman, in a burst of shared affection for their future offspring, putting her arm around the black woman. I've been musing whether the differing placement of the two commercials was coincidence or planning aforethought. Despite the frequent caveat voiced by knowledgeable people, that things often happen on television for no apparent reason, I'm inclined to believe there are no accidents. If Norman Lear can show blacks and whites embracing, why can't Pampers? But presumably the prime-time audience can accept it more readily.

I watched the prime-time commercial on a New York station, it must be admitted, and I don't know whether it was a national feed or a local spot, whether it played North, South, East, and West, or only in New York. There are a lot of black faces in commercials aired by New York stations that I suspect are less frequently shown, or not at all, in other areas, just as New Yorkers don't get a chance to see the ads for tractors, hog feeds, and herbicides that appear on Minnesota, Nebraska, or Iowa stations.

Commercials are one more factor in the division between the center (that is, the networks) and the localities that dominate the present and future of American television—perhaps, because they pay the freight, the most important factor. Television developed in the United States on a quantitative premise, that more was better, most was best. You wanted to put your programs and your commercial messages before as large an audience as possible.

On that basis, it was inevitable that the networks would take control of programming, and that local stations, putting their pocketbooks first, would go along. Local stations could make more money taking a program off the network line than by airing their own, since the networks could charge a higher per-minute rate by virtue of the larger audience they reached.

But in the past decade advertisers have learned to mix qualitative judgments into their strategies. It's not merely how many people you reach, but what kind. You want your commercial to be seen by people who are likely buyers of the product. Rating services like Nielsen and Arbitron offered stations and sponsors detailed breakdowns of audience composition by age, sex, income, education, and occupation.

This made "spot" advertising much more important—the buying of commercial time station by station, program by program (or at least time period by time period), rather than offering up the same ad to the entire country at the same moment. Let it be said that the networks are not hurting for advertisers. They still have less time to offer than buyers want to buy. But "spot television" is growing faster than any other print or broadcast form of advertising, increasing by 31 percent for 1976 over 1975, according to the Television Bureau of Advertising. This can only strengthen the hands of local stations in their conflict with the networks—over the expansion of network news to one hour, for example, which the affiliates managed to forestall, or greater clearance (advance screening) time for controversial programs.

So commercials are more than an arena for domestic psychodramas and sexual fantasies and half-minute comedies that delight a knowing student audience. They are the leverage that often determines what goes on commercial television *between* the commercials. We've seen the bandwagon spectacle of dozens of advertisers responding to public pressure by declaring their unwillingness to sponsor "violent" programs. If the end result were one or two fewer punch-outs per "Starsky and Hutch" episode, few of us would be the worse for it. But what happens when pressure-group tactics are applied to sponsors of potentially controversial documentaries? The most notorious incident of recent years occurred when most of the sponsors pulled out of the 1975 CBS documentary, "The Guns of Autumn," in response to special interest group demands.

CBS kept that program on the air—and even reported on the behind-the-scenes maneuvering of pressure groups and sponsors in a follow-up report. Documentaries, however, rarely do well in the ratings, and there is no reason to count on the value of prestige or principle outweighing considerations of profit and loss.

Behind those thirty-second stories may lie the most important story for commercial television today. It's one we're likely to miss when "there's a break in the action" and we head off for the refrigerator or bathroom. The one thing viewers pay less attention to than commercials is what commercials mean for the rest of television they watch.

5. Just Don't Show the Blood

Sgt. Hutch grabs the villain by the collar and hurls him headlong toward a stack of supermarket cartons. We see the man in motion, hear the thud of his head at impact, see the bad guy sprawled senseless. Wait, I must have blinked. I didn't *see* the guy actually hit the cartons and land on the floor.

I didn't blink. "Starsky and Hutch" didn't show the image. If you want to know what influence the past year's agitation against television violence has had on network television, the answer is: the jump cut. You get the beginning of a violent action, then its consequences. The precise moment of violence is excised. Moral crusades like the war on television violence are likely to have unplanned consequences, and the result so far seems to be a sudden turn in television melodrama toward an almost avant-garde narrative discontinuity.

That isn't the only consequence, of course. Since the critics are standing by, ready to tote up every depiction of violence, there is also less showing of violence and more talking about it. A man walks into a room and says, they did this to me and that to me. Finally, we see him in close-up, and his eye is blackened, his forehead bruised. Action-adventure shows are becoming positively classical in shifting the violence offstage, with messengers showing up from time to time to tell us graphically about it.

Not less violence but more distance from violence—that seems to be the response of networks and producers to the challenge of their critics. A villain holds a gun to the head of a grocer, but we see it in long-shot, small and faraway through the store window. A woman is kidnapped and held hostage, but we only know it because the bad guys show her ring to her husband. No one ever accused those who deplore television violence of being aesthetes—they're too busy with their tally sheets—but the next debate on television content will have to pay some attention to aesthetics, particularly the aesthetics of distance. There are more ways to be violent than by showing the flash of a firing gun.

In the "Starsky and Hutch" episode, where the villain's head never makes visual contact with the stack of cartons, not a single shot is fired on screen. Yet the entire program revolves around a gunshot wound—the only evidence of which is a discreet band-aid on the forehead—that is slowly killing Starsky's girl friend. Of actual violence, the program is the soul of discretion. Nevertheless, it presents instance after instance of blood-curdling verbal violence. A madman is out to break Starsky by killing off those he loves. "He's going to wish he'd never been born," says Crazy George, and proceeds fatally to injure the woman, whose tearful death occupies much of the hour.

For my taste, give me a car chase and a few flying fists. Those who claim that their concern is the sensibilities of children ought to pay more attention to the way children, and indeed all of us, use television entertainment. In my experience children quickly learn the conventions of action-adventure shows and accept them for what they are—ways of telling a story through violent action. Much more disturbing is the presentation of psychopathic behavior and incipient or offscreen violence. Here the conventions are murkier, the psychological reverberations less well controlled in the production, and the opportunities for excess more difficult to avoid. You can get up from some television shows, even though they are filled with violent acts, satisfied with having seen an entertaining fiction. Shows with little or no overt violence, but with violence of the distanced verbal kind, often perpetrate their threats, through their imaginative gaps and weaknesses, directly on the audience.

Television, as it is practiced by the American commercial networks, is as much a verbal as a visual medium. This is not to say that the images do not convey meaning and have pertinent aesthetic and iconographic value, but that networks and producers choose to emphasize the word. They sense, perhaps, that they are as likely to capture their audience through the sound track as through the picture. We might be some distance from the image, across the room or with our backs turned, doing some household chore. The picture our set provides may be less than perfect. We remain, despite the heralded twentieth-century triumph of the visual, a culture with a Puritan heritage, enamored of the word.

Nowhere is the domination of the word more evident than in prime-time television's treatment of sex, the other great frontier that television's watchdogs have not yet stooped to conquer. Overt sexuality of the kind that is now common in major feature (not just pornographic) motion pictures, ranging from four-letter words to nudity to scenes of simulated copulation, is rare to nonexistent in the regular weekly television series. Yet sit-coms are suffused with references to sex of the most suggestive sort. The sit-coms, in fact, seem to have turned into a nonstop Feydeau farce.

Though network censors have insisted in the past that Middle America worried more about sex than violence, the current mood seems precisely the opposite: Sex play is taken for granted far more than gunplay. And why not? Nobody gets hurt, not even emotionally. It's safer to play with fire on television if it's fire in the blood, not from a weapon.

The truth may be that American morality—at least official morality—knows its mind better about sex than about violence. Violence can be a relative thing. Guns fired by policemen or GIs may have a different moral connotation than do guns fired by thugs or Nazis. With the exception of those committed to nonviolence, nearly everyone has to ponder the specific situation before deciding whether acts of violence should be accepted or condemned.

On sex the official line brooks no such ambiguities. Sex is no longer reserved solely for the marriage bed, but if you're married you're not supposed to wander. More is believed to be at stake in marital fidelity than mere personal values. In a tryout episode of "Tabitha," a mid-season replacement series on ABC, that benign witch attempts to deter a married man from an affair, and she exclaims, "We're trying to preserve home, family, and the American Way." That line was read, as I viewed it, with not an ounce of irony.

Tabitha is a teenage version of Glinda, the good witch in *The Wizard of Oz,* but she seems to have been posted to Sin City rather than Emerald City. In the marriage-saving episode, she is pitted against Portia, a "sleek, black, and shifty" witch who sets her designs on Tabitha's black boss, Marvin, more in the spirit of competition with Tabitha than out of heart's desire.

No one else in the cast seems to care much for Tabitha's protecting the American Way. Her Aunt Minerva, another witch, says, "There's nothing more boring than a moral marriage." And her egotistic colleague, Paul, puts in, "If you do something rotten, at least do it in style." These last remarks are uttered in a nightclub called the Dirty Disco. Finally, however, Tabitha's white magic wins out over Portia's black magic, and husband and wife are together at the end.

There may be more to this than mere entertainment. Perhaps it is a parable on the evils of temptation, on survival of the trials of the flesh. If so, however, the trials of the flesh didn't end with the show. Immediately after the moral fade-out, viewers were further tested by a commercial for Playtex "sexy, seamless" bras, with a thirty-second montage of young men ogling chesty young women.

There seems to be a surfeit of sexual wordplay on television, even on shows which appeal primarily to younger viewers. On "Fish," another ABC show, a recent episode centered on the interest shown by a wealthy divorced woman in Mike's "career." The sound track seemed to be dominated by young people's laughter, and, when Mike says, "Mrs. Wilson wants me to live with her," the sound track carried an arc of oohs and aahs. (Could there be audience cue cards with "ooh" and "aah" written on them?)

ABC is not the only network playing the sex game. It only seems to be doing it better than the others. On MTM Enterprises' "We've Got Each Other" (canceled midyear by CBS), the plain wife, Judy, in her rimless glasses, had little more to express to her pudgy husband, Stuart, in his hornrims, than her sexual desire. "No one's going to bother Stuart" on the Hawaii beaches, a character says in one episode, and Judy replies, "That's not quite true. I'm going to be all over him." Later, when they are alone, she tells Stuart, "I *am* going to be all over you, you know." Again, she to him, "I'm trying to turn you on." Yet again, Judy to Stuart, "You wanna neck?" Lucky Stuart. Unlucky CBS.

One likely drawback to "We've Got Each Other" was that its lovers were married. However sanctified marriage may be in official American morality, the dramatic portrayal of marital bliss seems to leave audiences cold. In the wake of "Roots" and the election of Jimmy Carter, ABC executives were proclaiming a revival of family values on television. But ABC's swift rise to the number one network has been built primarily on shows like "Charlie's Angels," "Laverne & Shirley," and "Happy Days," where the concept of family has, at best, been played down or so redefined as to have become something else. Or are the three Angels and Bosley actually the "family" of unseen Charlie?

Amidst all the talk of "family," ABC was putting its money on shows more closely attuned to America's popular taste for sexual fantasy, and to the limits of prurience on our federally regulated airwaves. One of these new shows, "Three's Company," became a smash hit; another, "Soap," became notorious.

After watching the pilot of "Three's Company," and later noting its success in the ratings, I attributed its popularity purely to prurient interest. The show's basic story line revolves around a young man moving in as roommate with two young women. In order to satisfy the landlord's sense of propriety, they convince him that the young man is gay. (Of course, he isn't, but nevertheless there is only verbal byplay, no sexual contact, among the roommates.) Meanwhile, the landlord's wife constantly complains about her lack of sexual satisfaction. That's the way the show began, and half a year later it hadn't moved beyond those situations.

And then there's "Soap." As times passes, the furor over "Soap" looks more and more like a manufactured event, designed to take the moralists' minds off violence. ABC, of course, risked losing some of its more squeamish affiliates, and thus damaging the show's potential for high ratings, but indications at year's end were that "Soap" had slithered its way into the everyday world of television with few remaining holdouts.

"Soap" *is* raunchier than most of the other sex-oriented situation comedies, but the moral boundaries are as murky as ever. It is certainly no more forward in its treatment of sex than "Mary Hartman, Mary Hartman." And one of the better things that can be said for "Soap" is that the kids who used to stay up for "Mary Hartman, Mary Hartman" can watch "Soap" and still get an hour's more sleep.

Perhaps the greatest problem for parents is not how to cope with children who understand the show's sexual jokes, but how to deal with those who don't. A recent episode opened with a scene in the police station. In the background a woman who has been arrested is saying, "I told him he could go around the world for $100. I'm a travel agent, for God's sake." My advice to parents is, jokes like that are best left unexplained.

"Soap" might as aptly be called "Susan Harris, Susan Harris." Harris created it, produces it, writes it, and plays the prostitute (no, she's *not* a travel agent) in the scene described above. As an example of the show's sexual humor, consider the following scene:

Danny is breaking into the home of the gangster Lefkowitz. In the dark, he crawls in the window of the daughter, Elaine. She turns on the light, grabs a bedside gun, and points it at Danny.

She: "Are you a rapist?"

He: "No."

She: "Oh." (Disappointment. Pause.) "Are you absolutely sure you're not a rapist?"

He: "I don't know what you want."

She: "Get your clothes off."

He: "What? Are you serious?"

She cocks the gun.

He: "I mean, under this kind of pressure, don't expect much."

That's all we get, though in a montage reprise under the closing credits we see Danny down to his underwear, the gun still cocked.

It is of some interest to note that *TV Guide*'s brief description of this particular episode tells prospective viewers that Danny "beds" the daughter.*

Being forced to perform under threat of a gun may be called mixing sex with violence. Considering how easy it is to mold those two different forms of human contact into one concept, "sexand-violence," the surprising thing is how infrequently the two actually go together on the same television program. Television has been much more cautious than movies in mixing those twin allures—or taboos, depending on your viewpoint. Perhaps the shorter format of television programming requires simpler story lines; perhaps it was a conscious decision based on a judgment of what the living room audience would accept, and in what form. Those made-for-television movies and televised theatrical features that have breached the division, like *Death Wish,* have been among the most controversial offerings in recent years.

But there are indications that "sexandviolence" is making some inroads into regular network programming. The most obvious, of course, was "Charlie's Angels," with three female private detectives using their sexual charms more often than not as a means of entrapping the bad guys. Sex for the Angels is almost exclusively a utilitarian thing, a tool in crime control. I can't recall ever seeing an episode where Sabrina, Kelly, or the third Angel (Jill when Farrah Fawcett-Majors played the role, Kris this year with Cheryl Ladd) gave any sign of going to bed other than alone. With Bosley, it's a different matter.

Plenty of pulchritude, plenty of violence, plenty of chastity—that's the "Charlie's Angels" formula. A recent episode, "Game, Set, Death," offered more than usual of the first two elements. It dealt with a series of attempted murders on a women's tennis tour. The villains seek to kill one player by scalding her in a shower room; another by slipping a rattlesnake into her picnic bag; a third by shooting at her with a high-powered rifle while she's practicing her serve. All these attempts are graphically presented, and a little terrifying, though none of the intended victims

* With familiarity, I later learned to appreciate "Soap" more, especially after spending a few days on the program's set while preparing to write about its director, Jay Sandrich. See Chapter 9, "Jay Sandrich: Directing Situation Comedy."

actually is harmed, with the exception of the rattlesnake victim (who, we are given to understand, will be cured).

In the one successful murder, the victim is discovered sitting upright, dead, though without signs of violence. Later we learn she died of "respiratory failure." How the bad guys accomplished this sanitized murder is never satisfactorily explained.

Motivation for evil, indeed, is one of the more vexing pitfalls in the path of television scriptwriters. Why people get murdered in real life—family or lovers' quarrels, muggings and robberies, drug dealing, organized crime, and the like—is not necessarily the stuff of which television melodrama is made. Most true-to-life murder victims are poor and outside the white middle-class mainstream. There are ethnic and racial pressure groups that television must keep in mind.

In the "Charlie's Angels" tennis tour episode, the villains are promoters of a particular tennis racquet that has been endorsed by one of the woman stars. In order to make sure their endorser wins top prize, they set out to eliminate her principal competition. This is how I understood it, but don't take my word for it. Understanding motives for evil is not what "Charlie's Angels" is about.

On another action-adventure show, "Switch," an episode that mixed sex with violence made even less effort to motivate the malice. Here a taxicab company operated by women is subjected to various malevolent acts. The villains turn out to be jewelers who need to get into an unused sewer underneath the cab company office to rob their own store so they can . . . get even with somebody's mother, I think. Whatever, it leads them to speak lines like, "Gotta get those dames out of there, even if some of 'em have to wind up in the morgue."

Those dames, meanwhile, are a pretty together bunch of feminists, who have a big poster on their office wall announcing a meeting in support of the Equal Rights Amendment. But they're also, pardon the expression, gorgeous. And darned if, after all the bombings and muggings are over and the crime solved, the pretty boss of the cabbies doesn't go home with private investigator Pete Ryan. The only woman who is a butt of the script's humor, by the way, is a flighty, callow girl learning to be an auto mechanic, who is given the unusual but oddly familiar name, Tabitha.

Now I suppose you want me to tell you what it all means, the sex and violence, the innuendo and circumlocution, the jump cuts and the distancing, the morality and immorality of it all. Sadly, I have little to say that you don't already know. This is American commercial television, circa 1978. Just as you have to put out the newspaper every day, even when there's no news, you have to fill so many hours of television every day. Much of prime-time television is like newspaper boiler plate, the stale stuff you use when you have nothing fresh to say.

Sex and violence may seem more prevalent on commercial television because no strong alternative, no other clear programming interest or perspective, exists as a counterweight. But for the same reason, sex and violence *are* more prevalent. They're filling the time while producers and the networks bide their time. For some, filling time is an end in itself. These are the exponents of the "one hundred million Americans can't be wrong" school of television analysis. In reply to them, keep an eye on the decline in homes using television. As of last fall, the prime-time dip was 3 percent from 1976. It will be interesting to see how commercial television fills its time if more and more Americans fill their time doing something other than watching network television.

6. The Mysterious World of the Psyche

This happened on a "Charlie's Angels" show: Kris, the newest Angel, has witnessed a murder, has been struck by the murderer's car, has wandered off in a daze. The other Angels are looking for her. So is the murderer. The killer runs into Kelly and Sabrina and gives them a false lead.

"We appreciate your help," says one of the girls.

"Nice to be appreciated," replies the killer.

Nice to be appreciated! That line jolted me out of my somnolent concern for those pretty ladies in distress. Nice to be appreciated? Are we going to learn that this poor man turned murderer because he was deprived of appreciation in his youth? Nothing of the sort. His psychology is of absolutely no concern in the show. He's just a thug—but a thug who somewhere along the way picked up a line of jargon from the psychologically hip.

Once alerted by that oddly dissonant line, in fact, I began picking up from all corners of the prime-time television landscape snatches of "psycho-babble," the new language of psychological awareness fostered by best-selling self-help books with titles like *I'm OK—You're OK* and *Looking Out for No. 1.* Psychology, it seems, is one of the prominent answers to the question, What will television programs be *about* now that they've withdrawn, like many of the rest of us, from the "social relevance" that was prevalent in the early seventies?

But it figures. Television follows the popularity charts from other media the way the Supreme Court, in the old political dictum, follows the election returns. Popular movies spawn television program ideas, so why shouldn't popular books, especially when there seem to be dozens of such books giving aid, comfort, and instruction in how to improve the self in the new era of the Me Generation. Add to this the well-known popularity of psychiatrists in Beverly Hills, and you have a television-creative community not only primed to draw on the words and phrases from pop psychology books, but somewhat experienced itself in their use.

That episode of "Charlie's Angels," which couldn't have cared less about the killer's psychological problems, was, it turned out, all about the psychological problems of Angel Kris. Stunned by the blow of the killer's car, she had stumbled into traffic and eventually saved herself by getting into a taxi. Next we see her sitting on a Southern California beach, having visions of a little blonde girl playing in the sand.

Kris is suffering from amnesia, and her surroundings are calling up memories of her past. A beachfront volleyball game makes her think of herself playing volleyball as a college coed. Hawaiian music on a portable radio reminds her of a luau party—these "memory" visions are presented as iris shots with soft-focus edges.

Three thugs wander along the beach, see her dazed condition, and try to steal her purse. She pulls her pistol and scares them off, but then she is startled at the sight of the gun, drops it, and runs off. Later she tells someone who has befriended her, "I think I've been in some angry places," and then she has visions of Charlie's other Angels, Kelly and Sabrina, and of herself shooting someone to death.

Finally she has visions of the events that started the program, and her memory is fully restored—just in time for her to be attacked by the murderer, and for the other Angels and their sidekick, Bosley, to show up and subdue the villain. Back at their office, communicating with the unseen Charlie by intercom, Kris says, "When I was wandering around out there I kept remembering my childhood. It was nice, it was sort of peaceful. I'd like to be back there, sort of."

"You're loved," Charlie's voice reassures her over the intercom.

Of course, thought I, putting my skills as amateur psychologist to work: The week prior to this particular show Farrah Fawcett-Majors had made a guest appearance on the program. Those psychologically hip writers and producers were telling us that the return of Jill had made Kris, her replacement, feel insecure, and that during her amnesiac meanderings she was questioning her role as a beautiful but tough lady detective. She was retreating into the innocence of childhood from the hard world of violence—those "angry places"—she had chosen to live in. For this episode, titled, significantly, "Angel on My Mind," melodrama became psychodrama: a withdrawal, a rescue, and a reaffirmation of Kris's place among the Angels.

Psychological awareness, of course, means conflict. It's one thing to read a book about *How To Be Your Own Best Friend,* but it's another to put its precepts into practice when you're dealing with some other guy who's trying to be *his* own top buddy. Inner conflict and how it affects one's relations to others is not something television has presented very well in the past. Given the conventions both of sit-coms and action-adventure series, people with psychological "problems" have tended to be either comic stereotypes or murderous crazies. There have been few examples of what we know to be true to life, that it is normal to feel psychological conflicts and that a good deal of our energies are spent consciously or unconsciously coping with them.

The significant exception has been "Family," and if prime-time television continues its journey into the mysterious world of the psyche, this program may be regarded as the Jacques Cousteau of the genre, the pathfinding explorer and pioneer. The wise parents and precocious son and daughters of the "Family" family are so serious, aware, and sensitive that they can talk like books and *still* sound convincing.

"Family" is a prime example on television of the special relationship a continuing series can build with its audience. With many popular series, audiences develop a sense of familiarity and a set of expectations based on learning characters and formulas; with a few series—"The Mary Tyler Moore Show" is perhaps the most notable case—viewers gain a sense of trust. "Family" is one such series. The show has proved that it can take risks with its material without abusing the audience or leaving matters in a muddle.

One "Family" episode centered on whether to sell the country farm. Kate the mother grew up there, and she hadn't gone back since her mother, Hattie, died. The farm is legally in son Willie's name, but the parents foot the bills, and they feel they can't afford it any more. On a family visit to the farm they contact a real estate agent without telling Willie. Being his own best friend, and looking out for number one, he doesn't feel OK and doesn't think they're OK, either. He's hopping mad.

His mother apologizes, but he's not mollified. "No, I don't really think you're sorry," he tells her. "You know what's troubling you. . . . What went wrong between you and Hattie? Maybe you think if the house goes, everything between you and Hattie will be gone, too. Well, you're wrong."

If Willie had actually read any of those self-help psychology books, he might have learned that it's bad form to tell another person what you think is going on inside her head—not to speak of bad dramatic construction. Matters aren't helped much when Willie and his father discuss Kate's difficulties like two group-therapy veterans on a busman's holiday. The father speaks of "this terribly unresolved feeling Kate has about Hattie." The son replies, "I hope she can work it out."

These errors of commission can be forgiven in view of the larger aims, and large risks, of the program itself—to show how a woman's feelings about her mother color her relationship to her own children and to build the dramatic conflict among the living upon a psychological conflict with a person whom we, the audience, do not know or see. The strength of this subject, its capacity to touch anyone in the audience who feels unfinished business with a parent (and that's just about everyone), carries the program along. Willie apologizes to his mother for his tendentious remarks, Kate is shown, by means of voice-over, talking with her deceased mother in her mind, and they decide to keep the farm.

"Eight Is Enough" is closer in tone and intention to "Family" than any other series on television, but it is not yet close to the point of taking the risks with psychological subject matter that "Family" does. Not only is it still in bondage to a laugh track, it hides its psychological subject matter as a "back story" behind a forgettable, standard sit-com plot line.

To understand the episode I'm going to write about, you should know something about the history of this series. It began as an adaptation of newspaper columnist Tom Braden's book about his family of eight children. After a short tryout series was aired in the spring of 1977, the actress who played the columnist's wife suddenly died. A decision was made to launch the program the following season with the columnist, played by Dick Van Patten as Tom Bradford, a widower. Subsequently, however, he remarries, and the psychological incident in question concerns a daughter and his new mother-in-law.

This particular encounter fits into the prime-time convention of the morality lesson—one of the staples of the "social relevance" sit-coms of the early seventies. Don't be a racist, keep your money in a bank, respect the elderly, don't abuse your credit card: From lofty ethics to everyday advice, the sit-com plot seemed to point a moral finger at the audience, saying (with a smile and a laugh), Hey, you, watch out, shape up. In this case, however, it's not so much a moral as a psychological lesson.

Teenage daughter Elizabeth feels she needs a new nose. This, you'll concede, is not meaty enough for an hour of prime-time television, but it's where the emotions lie underneath—up there on the surface the family is rushing around preparing for a visit from the Vice President of the United States. Her beau has broken several dates, and she's convinced her ugly nose is the reason.

Nobody in the family much takes her seriously, and, of course, there's no money around for cosmetic surgery. But lo and behold, the stepmother's mother arrives, hears the tale of woe, and agrees to pay for the operation.

Now, even someone who never got past the preface of *I'm OK—You're OK* can understand what's wrong with this one. In her desire to be accepted as a grandmother, the woman has acted without giving sufficient attention to the girl's actual needs or feelings. Moreover, she's usurped the parents' authority. The stepmother and her mother argue out these points in an unusually realistic scene of domestic conflict (amid the baloney of the Vice President's visit). Finally the older woman sees the light: "I guess there is a fine line between trying to help and butting in."

This is soon driven home in a scene where Elizabeth bumps into her beau at a grocery store. It turns out that he has allergies that give him a facial rash, and out of his own embarrassment he's canceled their dates. As for her features, he think's she's perfect, nose included. Another little psychological lesson: The other person's behavior is as often based on what he thinks about himself as on what he thinks about you.

The people who live inside the television box seem unusually capable of giving each other psychological advice and counsel. You don't see them turning often to professionals, as they would, most likely, if their cars needed a tune-up. The practice of psychiatry and psychotherapy has received an ambivalent reception in popular entertainment ever since *The Cabinet of Dr. Caligari,* and television hasn't done much to alter the stereotype of the "shrink" crazier than his patients. Better to try and heal thyself than fall into the hands of one of those whackos.

This stereotype was paraded in all its hoary glory on a recent "Barney Miller" episode. The New York cops who inhabit that fictional station house are called upon to cope with three patients who escape from a private psychiatric hospital and make a disturbance in an Automat (maybe the person who wrote that hasn't been in New York lately; for verisimilitude he would have had to make it a McDonald's). Eventually the three escapees arrive at the police station, along with the loony doctor, a sex-crazed nurse, and a burly attendant.

The patients are a young man who had taken off his clothes in the Automat ("Nakedness is truth," he exclaims), an old man whose son has had him committed to get him out of the way, and a woman who thinks she murdered the husband who left her. There's material here for some poignant comment on mental institutions, but the intention of the program is not to go beyond preposterous low comedy at the patients' expense—and the psychiatrist's.

Dr. Engels, the shrink, is pictured as a fastidious prig who uses language as a shield against reality. When he arrives at the station house, he's upset at how the police talk about his missing patients. "Don't say 'escaped,' " he instructs, "say 'wandered off.' " He tells the cops that the patients won't be dangerous. "At worst," he says, "they're recalcitrant children."

But when the old man says, "Engels is out of his mind, and the place is a dump," we're likely, having seen Engels, the sex-obsessed nurse, and the goon attendant, to agree. "The first rule of psychiatry," Engels says superciliously in reply to the old man's remarks, "is not to reinforce deviant behavior."

"Sit on it, Engels," retorts the old man, and he seems to speak for the "Barney Miller" show as well as for himself.

Notwithstanding "The Bob Newhart Show," the standard sit-com calculation seems to be that more viewers will be pleased than offended by a put-down of psychiatry. But a large section of the population—though not perhaps the most avid sit-com watchers—regards psychiatrists and psychotherapists as helpers, healers, and heroes of our time, and television has begun to respond to this attitude, if not in standard programming, then in specials. In 1977, "Sybil" presented Joanne Woodward as a heroic psychiatrist who helps a young woman discover her true identity out of a welter of more than a dozen separate personalities. More recently, ABC has given us Tony Lo Bianco as a larger-than-life psychiatrist in "A Last Cry for Help," a television movie about teenage suicide.

"A Last Cry for Help," is, in the jargon of television sociology, a "prosocial" program, as opposed to what people who use this kind of language consider to be antisocial program content. It aims to build social values instead of (inadvertently) tearing them down. It's the story of a teenage girl, with a denying, misunderstanding mother and a passive father, who tries to kill herself and by her failure falls into the helping hands of psychiatry. (Curiously, one of the first things the psychiatrist says about the suicide attempt is that it's not a cry for attention, it's a real effort to end life, thus contradicting the title.)

So prosocial is "A Last Cry for Help," in fact, that it's like one of those comic books you used to get in school promoting, say, the virtues of good dental hygiene. Written and directed by Hal Sitowitz, who also served as executive producer, the television movie promotes the virtues of mental hygiene. This is not to say that it's unworthy or doesn't leave a lump in your throat at the end, only that its strength lies in the explicit way it asserts the case for psychotherapy. It's a message picture, though not from Western Union.

The psychiatrist makes his appearance midway through the film, after the high school girl has taken an overdose of sleeping pills and has ended up in a hospital. As played by Lo Bianco, Dr. Ben Abbot is like Columbo with an M.D., a soft-spoken, straight-talking, street-wise guy who knows the score. He knows what lurks in the hearts of men and women, knows that it isn't necessarily evil, just an unwillingness to recognize the truth about oneself.

Dr. Ben Abbot goes to work on the victim, Sharon, and has to deal with her parents as well, the uptight Mrs. Muir and her browbeaten husband. (Shirley Jones plays the mother, Linda Purl the daughter, Murray Hamilton the father.) "As parents you're not alone in this," he tells them, and when the mother resists recognition of the problem, he admonishes her, "Mrs. Muir, saving face is not the most important thing."

In her therapy sessions with the doctor, Sharon admits she hates her life because of all the things she is supposed to do and be for other people. Clearly she doesn't know how to be her own best friend, but Dr. Abbot can point the way. "I think you should have more control over your life," he says. If the film were a comic book, Sharon's response would be indicated by a light bulb over her head. That, her look indicates, is a new idea.

"Nobody wants to die," the psychiatrist says as Sharon leaves the hospital. "There are miracles—there are people who know and want to help you." To the defensive mother he warns, "She's not dealing with her feelings of hurt and depression." The mother asks what he wants of her. "Spend time with her. Listen to her." And sign up for psychotherapy, both of you.

But when they get home, the mother tells her daughter, "Nothing's changed. We still love you. Everything's going to be just the way it was." At that moment one might have heard a few million groans across America. Sharon, thank goodness, has the new-found strength to seek out Dr. Abbot at the suicide prevention center, and before long the daughter and parents are in therapy. Sharon gets a chance to tell her mother, "You're always telling me how to live my life, how to feel. Sometimes when I'm depressed or unhappy, I'm afraid to show it."

In the hospital Sharon had met another attempted suicide, a college student named Jeff whose mother had taken her own life and whose father is a domineering martinet. In her travail over her parents' failure to give her what she needs, she calls him, and they meet. But he is too far gone to help her or himself, and on his way back to school he drives his car off a cliff—an automobile "accident" that, as the psychiatrist has already warned us, is second to suicide as a cause of death among teenagers.

News of Jeff's death drives Sharon to renewed despair. She runs to the kitchen knife drawer, and she and her mother wrestle over the knives. Suddenly there is an emotional breakthrough. Mother hugs daughter and admits, "Maybe I don't know what love is. You wanted so much you frightened me."

Sharon returns to the suicide prevention center, and Dr. Abbot gives her the film's final message. She can't exist as the reflection of someone else's feelings, he says, whether of parents or boyfriend. "You have to be your own person." At the end, Sharon's family seems to be on the way to becoming as psychologically savvy as the "Family" family.

What will become of American culture if television moves further in the direction of psychological wisdom I'm not prepared even to imagine. Television reaches far more people than the most popular best-sellers, and there's no doubt that we adopt some of the ways of speaking and behaving that we observe in mass entertainment. A nation of people talking like therapists could become a nightmare of mulled motivations, pondered psyches, and ruminated relationships. Still, as they say, it's nice to be appreciated.

7. Sit-com Squabbles: Love by Other Means

I have been seeing more love taps on prime-time television lately. Love taps are those half-playful, half-serious little gestures of mock violence—the gentle chuck under the chin, the pulled punch to the shoulder—that wordlessly say, "It's only love that holds me back from bashing you."

Mock violence, both in words and gestures, seems to be on the rise in situation comedies these days. Since I'm not a social scientist, I haven't got any graphs or mock violence counts to prove it. It's just a feeling. Maybe it depends on what one is looking for. I set out to explore how people talk to each other on television, and what I found night after night was anger and contentiousness. Those love taps, to paraphrase the Clausewitz military dictum, were merely an extension of argument by other means.

Of course, it's nothing new to see loved ones shout and scream at each other in sit-coms. Any weeknight around midnight I can tune in "The Honeymooners," that classic of the fifties, and watch Ralph Kramden blow his stack. But the blustering bus driver serves as no model for the present-day sit-com. Nowadays characters are a good deal cooler about their feelings. The point is not to vent your wrath completely—hence those restrained love taps—but to win the battle of verbal wit.

Before expounding on the dire consequences of detached witticisms for modern civilization, I fortunately chanced upon a book that opened up another perspective. The famous British study, Iona and Peter Opie's *The Lore and Language of Schoolchildren,* found that after the rhymes and songs of games the most important element of language play for children was repartee—the spontaneous one-upmanship of being more clever than the next guy. The put-down, it appears, far from being an electronic age invention, was splitting sides with laughter as far back as good Queen Bess's time.

As in so many other instances, television is less a leader than a follower. The medium incorporates elements from the larger culture that surrounds it. In this case, television these days seems more and more to be emphasizing verbal styles favored by adolescents of all ages—the cheeky impertinence that makes it better to be witty than to be right; indeed, that makes being witty being right.

Nonetheless, having acknowledged that there is nothing new under the sun and that television may not be saying anything to us we're not already saying to ourselves, it's important to observe what the power of the medium can do. Television can establish a context, and it can create an aura. The aura can arise from constant repetition: Everybody's doing it, night after night, show after show. Sit-com characters spend so much of their time these days trading one-liners that when they actually hold conversations, one might mistake it for Shakespeare.

The context develops out of who's saying what to whom. A year or two ago on prime-time television contentiousness seemed to be directed outward, toward snooty salespeople, slumming socialites, haughty intellectuals, and other antagonists across the boundaries of status and class. More recently, the arguments seem to have moved closer to home, to within the intimate family circle, real or surrogate. And it has become harder to understand what the arguments are about any more, other than argument for the sake of argument: in other words, for the convenience of plot construction.

Take an episode of "Three's Company." It featured not only love taps, if that's a proper word to use among nonconjugal roommates, but also downright knocks and punches, as well as a veritable war of words—and all in service of an arguing point: Which sex has the greater willpower?

It seems that Chrissy has been stuffing her face with food more than usual and Jack has been chasing—and catching—a luscious blonde who works on a cruise ship. The *TV Guide* description of the show says, curiously, that "to test the willpower of men versus women, Chrissy begs off food while Jack abstains from romance." But, in fact, it isn't like that at all.

What actually happens is that Janet, the third member of the ménage, forces her two roommates to swear not to touch what it is they most desire. Chrissy doesn't swear off food, she lusts after it; Jack doesn't willingly abstain from romance, he actively craves it. But both succumb to the concocted argument about willpower, until the show degenerates into screams of thwarted desire and a final shot of Janet pushing a piece of cake into Chrissy's face.

Did the thirty or so million people who watched this show know what was going on? I think I do. I think the real story of that episode is Janet's longing to be romantically involved with Jack and to be as slim and pretty as Chrissy. She's out to frustrate their desires because she's frustrated in her own. The abstract argument that Janet invents is simply a sublimation for her own more fundamental arguments with her roommates. How can Jack be interested in other women and not in her? How can Chrissy eat so much and still keep her enviable figure?

Wittingly or unwittingly, intended or by coincidence, ABC's powerful Tuesday night spring lineup, of which "Three's Company" was the centerpiece, was crowded with contentiousness. When characters weren't actually engaged in arguments, real or phony, they still seemed armed and ready for one. They're not about to take any guff. They're looking feisty, they've got a chip on the shoulder, and they're aching for someone to knock it off.

Shirley is baking cookies for a church social on "Laverne & Shirley," and Laverne sneaks one off the cookie sheet. "Get your stinkin' hands off those cookies," growls Shirley, usually the sugar-and-spice-and-everything-nice of the twosome. In fact, the nub of this particular episode has to do with Shirley coaching Laverne on how to be more of a "real woman"—that is, not so full of her usual vinegar and bile—so she can spark a romantic interest in her friend Joe, who sees her only as a companion for outdoor rather than indoor sports.

Joe comes over on a Saturday morning dressed in his rubber pants, planning to go fishing with Laverne, and finds her decked out like Scarlett O'Hara, preparing to serve him tea in eggshell-thin cups. Naturally, he looks at her in a different way—who wouldn't?—but as their tête-à-tête progresses, it gradually becomes clear that Laverne would rather win an argument from Joe than win Joe.

When Joe falls for this new image—a "real woman" but, of course, a false Laverne—she suddenly grows angry at the thought that her true self had been so ignored and devalued. She rises up in indignation from the tea service that is alien to both of them.

"Maybe the next time you should take it with milk," she says, pouring the contents of the cream pitcher down Joe's rubber pants. As he bolts out the door, she yells after him, "And grown men shouldn't wear rubber pants."

Perhaps you've noticed from the dialogue I've quoted that these shows, which are among the top five in audience popularity, wouldn't rank in the top fifty for the cleverness of their put-downs. That may be a sign of the difficulty of being original. The best put-downs, though spontaneous in the moment they are uttered, tend, as *The Lore and Language of Schoolchildren* suggests, to fall into well-defined patterns and to have been perfected after literally centuries of youthful use.

"Happy Days" this spring had a reprise show of some of its better moments, built around the pretext that Mork of Ork (and now of "Mork & Mindy") returns to the Cunningham household, where he first alighted on Earth, to learn more about how humans conduct their personal relations. It turns out to be almost an anthology of arguments, or, to be more precise, fights, near fights, and plain insolence.

The show reminded us, if we needed it to, that Fonzie is truly the grand panjandrum of put-downs. (Note: Panjandrum is itself a put-down word, a mock title said to have been coined in the early 18th century.) His special skill is not to fight but rather to end fights before they begin, not only by his threat of retaliatory force but also by the right choice of crushing words. To one hood, who is manhandling one of Fonzie's middle-class friends, "I would let him go, unless you want to make medical history." To another greaser with a gang. "Well, if it ain't the turkey and his feathers."

One more excerpt from the "Happy Days" archive: Father Howard Cunningham is standing on a table being fitted with a toga for a costume party.

"I'm Julius Caesar," he tells his daughter. "What do you think of it?"

"No wonder they stabbed him," she replies.

Not to get too technical, but *The Lore and Language of Schoolchildren* makes an intriguing distinction between repartee and guile. Repartee is a spontaneous response, guile is malice aforethought. Sit-com put-downs do fall into both categories—in the War Between the Ropers, which forms the back story of "Three's Company," they can both be found, separately or mixed—but the guileful one is as often the victim as the perpetrator of a put-down.

And it's a rare moment when you find spontaneous guile—that is, a character suddenly deciding to set up another for a put-down—but it happened in a "Delta House" episode. Otter, sitting on a sofa with an arm around a girl friend, says, "I'm alone, all alone."

"What am I?" she replies indignantly. "A throw pillow?"

That's not a put-down, of course. That's self-defense. The put-down is forming in her mind.

"Otter," she says, "you have no idea what it means to be truly alone." (A pregnant pause.) "But now you're going to see it." And she splits.

With characters like Fonzie and Laverne, contentiousness turns out to be a style, and argument a device. They aren't really contentious. It's just a tic or trait of their social masks, for deep down both are as sweet and warmhearted as the next guy or gal. Their feisty characteristics seem intended to add a little bite to what would otherwise be fairly bland shows.

"Happy Days," and "Laverne & Shirley" are products of the production unit at Paramount Television headed by Garry Marshall. These nostalgic and youth-oriented shows climbed to the top of the ratings in the latter half of the seventies, replacing as bellwethers of sit-com formulas the contemporary topical comedies of Norman Lear.

The Lear companies, of course, are still among the most prolific producers of situation comedies. Though they are not oblivious to the recent sit-com trends, they manage to incorporate new developments into the distinctive Learian manner. On the Lear shows, lessons still lurk behind the fun. If prime-time television is filled with contentiousness and argument these days, the Lear programs are likely to enlighten us on the meaning and implications of Contentiousness and Argument.

Those very topics, indeed, were on the syllabus during recent episodes of two Lear shows. On "One Day at a Time," two young filmmakers arrive at the free clinic where Julie (the daughter played by Mackenzie Phillips) works. They're going to shoot a documentary. Julie immediately starts squabbling with the director. In no time they're hurling insults at each other in a rapid-fire pattern of put-downs that most other shows couldn't match in a month. Their feelings toward each other are characterized, as one of them says, by mutual dislike, general animosity, and common suspicion.

But we know right from the start that the insults and the fights are just a ruse. As with Fonzie and Laverne, contentiousness masks real feelings. But while on those Garry Marshall shows contentiousness is a plot device, on this Norman Lear program Contentiousness is the subject itself. Julie confesses to her mother her attraction toward Nick the filmmaker. Their similarities of character, she says, enable him to know all about her, so her feistiness is a way of controlling her own feelings as well as hiding them from others. (The same is true for him.)

Wise mother Ann—after throwing her apron at Julie, another version of the love tap—convinces her daughter in a quiet talk that it's better to let her feelings out, to express them and experience them. More harm may come from their suppression. So Julie admits to Nick her romantic feelings toward him, and he reciprocates.

"I don't want to fight any more," she says.

"What's the matter?" he replies. "Don't you feel well?"

An episode of "The Jeffersons" carried further the notion that contentiousness and argument are really surrogates for something else. Where George Jefferson is concerned, you might think that, for once, a television character is snappish because his true self, deep down beneath all the masks, is also snappish; but no, even here there is a ray of hope for other, better selves.

The theme of this episode is Argument. Tom and his son Allan, two of the white characters on the show, are having a dandy one, hollering at the top of their lungs, trading insults that bear no relation to reality. George and Louise Jefferson, husband and wife, decide to stage a phony argument as an object lesson to Allan that when people argue they often say things they don't really mean.

Naturally, George and Louise's staged argument quickly turns into a real one, and what these two characters chiefly demonstrate is how adults can't argue like adults. (The dialogue in this program, curiously, is completely different from that of "One Day at a Time." Instead of zappy one-liner and zinger put-downs, the humor is entirely situational.)

The situation rapidly falls apart, however, and it is only rescued by the arrival of a deus ex machina in the form of a reporter from a black magazine who is supposed to interview George for an article on successful black men. This chap, it turns out, is a hearty advocate of the therapeutic value of arguing.

"Are you sure it's good to argue?" Allan asks incredulously.

"When a person takes the time to argue with you," replies the improbable reporter, "it means they care about you."

"Gee," says Allan, "my dad must really like me."

So argument, too, is a mask for caring, or at least the desire to care. If those love taps to the jaw were an extension of argument by other means, then argument itself, in turn, is an extension of love by other means. We've come a long way in a short time; it wasn't so long ago on prime-time television that sit-com characters argued *about* something, the rights or wrongs of work or war. Now when they're fighting, what they're really trying to do is express love.

It may be that the Garry Marshall unit at Paramount once again had its fingers on the pulse of the times when it gave us the instant hit of the 1978–79 television season, "Mork & Mindy." For Mork of Ork, played by Robin Williams, is the very antithesis of argument and contentiousness. Mork is Mr. Congeniality himself. Mork takes everything in stride and with a smile.

One of the touchstones of the show's humor is how different Earth people are from Orkans, and, indeed, Mork carries accommodation so far he turns it into parody. Nobody *human* could be as genial in the face of adversity as Mork is.

Does Mork's demeanor portend that the wave of contentiousness that is now running rampant on prime-time television will soon be over—that argument, the curious surrogate for love, will be replaced by love itself? Not likely. For Mork himself is beginning to learn the ways of Earthlings. I have not yet seen him arguing with a human being, but he has been heard muttering argument monologues with himself. ("Can you spare a word, Mac?" "Shut up." "Give me a chance.")

On sit-coms, love makes better ideology than plot device. On this planet, at least, we'll see more mock-violent love taps, verbal fights, and put-downs than true love.

PART TWO

The People and Practice of Television

8. Cold War in Televisionland

Not more than two minutes after I sit down with Frank Price, president of Universal Television, his telephone rings. Price doesn't take calls during interviews. This caller must be someone special. It is. Fred Silverman, president of ABC Entertainment, titan among network programmers, is on the line.

They have a pressing problem—a complication in casting the pilot of "Operation Petticoat," a half-hour sit-com based on the 1959 movie. Universal is producing the show, and ABC will air it Saturdays beginning in the fall. Names are bandied, meetings are planned. The call ends, and Price resumes the interview. "When it gets down to two days before it shoots," he explains, "we get a little nervous."

Frank Price does not look the nervous type. He's being polite. Or perhaps he wants to divert my attention from the facts of television life—even for Universal, titan among producers. When the network calls, the call goes through. In the television world, the networks call the shots. Producers are like farmers who need somebody else to transport their crops to distant markets. Only in television the middlemen, the networks, never buy the whole crop.

For television producers, this dependency upon the networks feels more these days like a hair shirt than a silk blouse. They make, the networks take . . . or leave. Since the networks got almost completely out of producing their own entertainment shows a few years ago, and remain out under strong antitrust pressure from the Justice Department, they need the producers as much as the producers need them. But the networks still set the rules of the marketplace.

A few production companies are beginning to try to change the balance they see favoring the networks. Norman Lear has been rebuffed in one such recent attempt, but others have been more successful. Lear failed to gain CBS's agreement on a proposed rising scale of payments for his show, "A Year at the Top," should it draw an audience share higher than a stipulated figure. However, Grant Tinker did place with CBS an MTM Enterprises program, "Lou Grant," on an "on-air" basis, without prior network approval of a pilot. And Universal launched Operation Prime Time, a special network of independent stations, as an outlet for its multipartite "Testimony of Two Men," an adaptation of the Taylor Caldwell novel.

The producers' search for greater leverage with the networks may have been one of the reasons, along with falling ratings, why CBS decided to move its programming department to Los Angeles from New York. Neither ABC nor NBC appears ready to make the transcontinental leap, however. "You must not get too close to the creative process," Fred Silverman was quoted in *Broadcasting* magazine. "If you do, you lose your objectivity, what they call your third eye—the ability to stand back and see objectively."

On the morning Silverman called Frank Price, was he watching "Operation Petticoat" with his third eye or his two subjective eyes? One eye or two, it's a scrutiny under which some outspoken producers are openly chafing. Not only do networks have the ultimate voice in what goes on the air, but when a show succeeds, the network, rather than its producer, gets the credit. The public at large recognizes "Baretta" as an ABC show, "Kojak" as a CBS show, "The Rockford Files" as an NBC show, and hardly seems aware that all three shows are products of Universal City.

One of the few instances of production companies working in concert came when a group of producers brought suit against the "family viewing" period, a policy adopted by the networks, under pressure from the Federal Communications Commission, that restricted programming subject matter during the first prime-time hour (the aim was to curtail sex, violence, and "adult" subjects during a period when children's viewing was supposedly at the highest). The producers won a victory when a federal judge ruled that the network restrictions violated the First Amendment guarantee of free speech. "The dogs are barking at the heels of the networks," says Grant Tinker, one of those producers who doesn't mince words. "The heels of the networks—there are a lot of them."*

Maybe Fred Silverman called Frank Price on a hot line. Sometimes the invective you hear makes the relation between networks and producers sound like a Cold War.

As head of the largest television production company in Hollywood, Price is too circumspect to join in the battle of words. At Universal Tower the image and demeanor seem corporate, more like Sixth Avenue than Sunset Boulevard. Other West Coast television executives wear sweaters and open-neck shirts; at Universal they're in suits and ties.

If Universal is having troubles with the networks, Price doesn't need to advertise them, any more than he needs to crow about his company's success. Universal Television is to television what Universal Pictures' *Jaws* is to movies—the biggest thing around. For the upcoming season Universal is producing thirteen regularly scheduled shows that will occupy twelve hours of network prime time weekly. No other production company is responsible for even half as many. Add in all the mini-series Universal will put on the air and you will get some weeks when fully one-fourth of all prime-time network programming is provided by Universal. There may be nights—Wednesdays, Fridays, and Sundays are candidates—when at certain hours the network offerings will be universally Universal.

* Late in 1979 a federal appeals court overturned the decision won by the producers. The three-judge panel held that "the district court should not have thrust itself so hastily into the delicately balanced system of broadcast regulation," and ordered that the issues raised in the controversy over the "family viewing" period be returned to the Federal Communications Commission for further consideration. However, the networks have made no move to reinstate the restrictions of the "family viewing" concept.

This is the awesome empire over which Frank Price presides. A man in his middle forties, Price has been with Universal nearly twenty years, beginning as a writer and producer on "The Virginian." As producer of "The Doomsday Flight" for NBC's "World Premiere" program, he was instrumental in Universal's breakthrough into big-budget, made-for-television movies. And as head of Universal he has fostered its expanding commitment to "long-form" television. Price speaks of his relation to the networks in more discreet terms than some other producers, but behind his words you sense that Universal, no less than anyone else, has had its share of struggles.

"Much of the thinking of the networks came out of the old radio patterns," Price says. "The radio networks would get a hit and let it run five or ten years. You don't interrupt the schedule for anything, only certain specials over a period of years. There never was an interest in putting on a series with a built-in ending on it. It seemed rather pointless to the broadcasters."

For nearly a decade, Price says, Universal has been trying to alter the programming structure television took over from radio. "We are substantially responsible, if not totally responsible, for the development of the limited series, the novel for television, the mini-series"—he gives three different names for the same phenomenon. "The first successful attempt in that form was "Vanished," a four-hour, best-seller novel for television. That didn't kick off the trend, though. The breakthrough came with 'Rich Man, Poor Man.' It led to ABC going ahead on 'Roots.' "

As far as Price is concerned, it was the networks rather than audiences who were unwilling to consider changes in television's format. "I think the audience has always been ready," he says. In any event, "You can't find out from an audience what they want to see. You have to do something for an audience and see how they respond. If you have the right subject matter, you take the entire evening."

Price recalls an occasion when ABC devoted all its prime-time hours one evening to a documentary on Africa as the moment he became convinced that long-form television would work. If the networks believe a subject is important enough to take over an en-

CBS

The television icon of the 1970s was Archie Bunker's living-room chair, on "All in the Family." Norman Lear's situation comedy about the bigoted foreman from Queens, New York, was the top-rated prime-time show for five consecutive seasons, from 1971 through 1976. When "All in the Family" went off the air in 1979, the chair was donated to the Smithsonian Institution, storehouse of Americana. The chair, of course, faced a television set.

CBS

CBS

After "The Mary Tyler Moore Show" ended its run in 1978, characters from that popular program went their separate ways. Moore displayed her talents as a dancer in two short-lived variety programs, "Mary" and "The Mary Tyler Moore Hour" (top). Edward Asner continued as Lou Grant in a new series named for the character, who became city editor of a fictional Los Angeles newspaper; here (bottom) Grant faces the crisis of a city power failure along with publisher Mrs. Pynchon (Nancy Marchand).

CBS

Norman Lear pioneered in bringing black situation comedies to television in the 1970s, but by decade's end some blacks chafed at the comic stereotypes they perpetuated. Sherman Hemsley and Isabel Sanford played George and Louise Jefferson (top) in "The Jeffersons," a spin-off from "All in the Family." "Good Times" (bottom) derived from "Maude," which also originated with "All in the Family." Esther Rolle was Florida and Jimmy Walker played her son J.J., whose expression, "Dy-No-Mite," was much copied by young people.

CBS

ABC

ABC

The David Wolper television adaptation of Alex Haley's *Roots* drew a larger audience than any other series in the history of American television. Here (top) Kunta Kinte, Haley's ancestor, captured in Africa and sent to the American colonies as a slave, attempts an escape. LeVar Burton plays Kunta Kinte. Wolper followed "Roots" with "Roots: The Next Generation," carrying the tale of Alex Haley's family into the author's maturity. James Earl Jones played the mature Haley (bottom) interviewing Malcolm X (Al Freeman, Jr.).

NBC

In the aftermath of "Roots," some producers attempted to depict blacks on series television in more realistic ways. "Harris and Company," with former football star Bernie Casey in the lead role, was one such program and had a brief run on NBC. Casey as Harris, a tow-truck operator, here confers with Beverly Todd, a woman Harris dates.

PBS

PBS

Public television emulated prime-time sit-coms with "¿Que Pasa, U.S.A.?" (top) and tried to match British adaptations of literature for television with "The Scarlet Letter" (bottom). The situation comedy from station WPBT-Miami featured a Cuban émigré family with three generations under one roof. The adaptation of Nathaniel Hawthorne's novel starred Meg Foster as Hester Prynne and Kevin Conway as her husband Chillingsworth, here shown in a production shot from WGBH-Boston.

Don Perdue, WNET-New York

For intellectuals, public television station WNET-New York provided a weekly quiz game on the news, "We Interrupt This Week," and "The Dick Cavett Show," a nightly half-hour with the talkmaster and guests. The regular panelists on "We Interrupt This Week" included (top) author Barbara Howar, critic Jeff Greenfield, lyricist Peter Stone, and journalist Richard Reeves. Cavett (bottom) converses with guest Professor Paul Weiss.

© 1980 Don Perdue

NBC

A harbinger of television in the 1980s may be "Real People," created by George Schlatter for NBC, a program intended to display the wild and wacky ways of living human beings, rather than fictional characters. Here the show's co-hosts (from left) Byron Allen, Skip Stephenson, Sarah Purcell, and John Barbour applaud the antics of the San Diego Chicken.

tire evening on prime time, he suggests, then the audience would recognize its importance, too.

Now, with the old pattern broken, Price says, "There is greater flexibility toward the various forms, and that's to an incredible degree. Almost any kind of project can be undertaken and can be shown in the proper way. A network won't be frightened of that."

Here's one striking example of this flexibility: Price describes to me a long-form special, "What Really Happened to the Class of '65?," based on the nonfiction book by Michael Medved and David Wallechinsky. Not long afterward it turned up on the NBC fall schedule as a weekly, hour-long series.

Universal also has in the works multipartite productions based on two other important nonfiction books, Lucy Dawidowicz's historical study, *The War Against the Jews,* and Susan Brownmiller's book on rape, *Against Our Will.* Novels to be adapted for television include *Wheels, 79 Park Avenue,* and *Aspen.* And the most extensive project of all involves the re-creation for television of James Michener's novel, *Centennial,* in no fewer than twelve, two-hour movies.*

"The exciting thing about the novel form for television," says Price, "is that the television medium provides the proper outlet, for the first time in the history of drama, for this kind of material. It couldn't be twenty-four hours in the theater, as a play, or as a motion picture. Previously we did digests of the novel."

Meanwhile, Universal continues as the most prolific producer of regular television series. ABC will continue with "Nancy Drew/Hardy Boys Mysteries," "The Six Million Dollar Man," "Baretta," and add the subject of the hot line call, "Operation Petticoat." CBS retains "Kojak" and "Switch." NBC draws most heavily on Universal. It holds over two shows, "The Rockford Files" and "Quincy," takes over "The Bionic Woman" from ABC, and adds three new Universal offerings—"The Oregon Trail," an hour-long Western adventure series; "Rosetti and Ryan," a court-

* "Wheels," "79 Park Avenue," and "Aspen" appeared on network television during the 1977–78 television season, the twelve-part "Centennial" during 1978–79. Nothing more has been heard, however, about adaptations of the two non-fiction books. Frank Price, meanwhile, left Universal to become an executive of a motion picture production company.

room series about a pair of criminal lawyers; and "What Really Happened to the Class of '65?"

"The series will always be with us," Price says. "The basic thing is to come up with something distinctive. I think we have created a few trends there: the six-million-dollar man, the superhero. 'Emergency' has had a substantial impact—there will be more imitations of that. For a series you need a central character whose presence in the melodrama is legitimate, who has some credibility going for him. That's why detectives, cops, doctors, and lawyers make up 95 percent of television drama.

Price staunchly defends Universal's action-adventure series in the face of the continuing campaign against excessive violence on television. "I'm very proud of our police shows. Popular entertainment has always had some violence with it. There are two very compelling things to people—life and death. Life is directly tied to sex. Death is directly tied to violence. It's no real surprise an audience would want to watch shows dealing with these subjects."

Price emphasizes the restraints that operate in the making of a television show. "We have our own taste that we use in making our shows. The networks then apply their own standards. Ultimately, what comes out has been through a sieve.

"I would always be against censors. The net effects of free information far outweigh any negative effects that could come out of it. In a free society you will see some violation of good taste—that goes with a free society. In *Brave New World* all conflict and violence were eliminated from entertainment."

Frank Price believes in the positive values of television. "We're better off after twenty-five years of television," he says. "If it were not for television, we'd still be fighting in Vietnam—the South would not have changed so rapidly. If there were television around the world, it would be hard to have any wars."

That last remark rings familiar. Among my research notes I trace its analogue. In a book by Edward S. Van Zile, *That Marvel—The Movie: A Glance at Its Reckless Past, Its Promising Present, and Its Significant Future,* the movie medium is held forth as the hope of civilization to end war. That was in 1923. Since then there have been a lot of good movies and a few terrible wars.

Television would build a more convincing case for itself as an antidote to war if the men who make the programs and the men who run the networks could demonstrate greater success at peacemaking among themselves. A few miles west of Universal City, along the Ventura Freeway, lies Studio City, where Grant Tinker, president of MTM Enterprises, surveys the relations with combative but realistic candor.

"A lot of the producers think it's time to move in and attack through the Justice Department a lot of the practices the networks have been getting away with for years," Tinker says. "All of those things that work to make theirs a good business and ours not so good.

"My fear is that we might be out of business ourselves. The networks act as policemen in our forest, and without them there might be anarchy. Universal might come in and gobble us little guys up. I'd like to see the law of supply and demand work a little better. I don't want to see it change that much."

The law of supply and demand isn't working all that badly for MTM. Though MTM may be a "little guy" compared with Universal, the production company founded on "The Mary Tyler Moore Show" will have six regularly scheduled programs on the networks' fall schedules, making it second only to Universal in the number of shows on the networks. In addition, reruns of "The Mary Tyler Moore Show" will be in syndication, and Mary herself will perform in three specials during the coming year—an adaptation of *First, You Cry,* the book about breast cancer by Betty Rollin, and two musical comedy specials. MTM is also producing a twelve-hour version of the William Goldman novel, *Boys and Girls Together,* for NBC.

All six of the MTM series will be on CBS. Three are holdovers—"The Bob Newhart Show," "Rhoda," and "The Tony Randall Show," which moves to CBS from ABC. Of the newcomers, "Lou Grant" will be the most direct survivor of the breakup of the WJM-TV newsroom team. Edward Asner will continue to play the part of Lou Grant, who will move to Los Angeles and become a newspaper editor on the hour-long comedy-drama. "The Betty White Show" will be a half-hour situation comedy about a former movie actress building a new career in television. "We've Got Each Other," another sit-com, is about a marriage in which the husband stays home and the wife goes out to work.

For a little guy MTM displayed a lot of clout when it came time to negotiate "Lou Grant" with CBS. The network agreed to waive the normal procedure of withholding a scheduling decision until it could see a pilot program. It accepted the program on an "on-air" basis without ever seeing a pilot. "We just asked for and got a series commitment," Tinker says. "What's good about that is that the audience makes the decision."

Grant Tinker has been a network executive in his career—during the 1960s he was West Coast vice-president of programs for NBC and briefly returned to New York as NBC's vice-president in charge of programs—but he seems to put his trust in audience judgment of program quality more than in the evaluation of network executives. The classic case, of course, is MTM's own Mary. "The Mary Tyler Moore Show" began rather poorly by the conventional network methods of assessment. "The first Mary show tested very badly," Tinker says. "If it had been a pilot, it wouldn't have gotten on. I would rather take my chances with the audience. That's the way I'd like to go."

There's a certain caustic air in Tinker's remarks about network television that seems curious in light of his company's exceptional success. Perhaps it's his character—Tinker is an open, straightforward, blunt-speaking man, lean and athletic-looking, dressed in color-coordinated browns and yellows, wearing a V-neck pullover on a chilly California day. Perhaps it's his audience—this writer, whose critical reference to MTM's "The Tony Randall Show" in print elicited a few more caustic words. Perhaps it's the feeling that Norman Lear and other producers have expressed, that the networks take a disproportionate share of profits from the successes producers create.

"We're very bad at our business," says Tinker. "We're entirely on film," more expensive by far, of course, than videotape. "Here we're entirely on deficit."

But there's another leitmotiv that runs through Grant Tinker's conversation, and it suggests that his pique is directed not only at the networks generally, but specifically at one network, ABC. In putting together its fall schedule, ABC considered three MTM shows and opted for none of them. It dropped "The Tony Randall Show," but CBS picked that one up. And it failed to program two new MTM shows, "Bumpers" and "The Chopped Liver Brothers," now in the limbo of unsuccessful pilots.

"Bumpers" is a blue-collar comedy about a worker in an automobile factory. ABC gave it a sneak preview in the half-hour slot before the Ali-Evangelista heavyweight prizefight. Sneak preview indeed: *TV Guide* listed that time slot as "To Be Announced." Predictably, "Bumpers" scored an unimpressive rating. "The Chopped Liver Brothers" is a comedy about a struggling nightclub comedy team, written by and starring Tom Patchett and Jay Tarses, MTM writer-producers who were once a struggling nightclub comedy team.

Grant Tinker's troubles with ABC go back to before the reign of Fred Silverman. An MTM show of which he speaks with fond regrets, "The Texas Wheelers," failed after four Friday nights on ABC in 1974 when "the joke was that Patty Hearst was hiding on ABC Friday night." But ABC has gone from the bottom to the top during Silverman's tenure, and the current failure of MTM to make that network's 1977–78 season undoubtedly can be traced to Silverman's decisions.

Tinker obviously is not pleased that the two MTM shows ABC has under contract didn't get a chance to find an audience, and can't be offered to other networks. Of Silverman, he says, "He's a guy with a big appetite, he's a greedy kid." With Silverman putting money into development, Tinker says, ABC constantly has on hand more shows than it can program. ABC's high rate of program turnover last season, as well as its leadership in the shift to year-round introduction of new programs, can be traced, Tinker suggests, to the network's need to play off its surplus.

Meanwhile, MTM is moving ahead with new ideas. One is a family saga serial program that would be offered for late-night viewing, from 11:30 p.m. to midnight, Monday through Friday. (Norman Lear's "All That Glitters" went on during that half hour on independent stations, with weak ratings once the novelty of its sex role reversals wore off.) Tinker likens it to the old radio serial, "One Man's Family." It would be built around an older couple with five grown children, and begin in the early 1960s, around the time of President John F. Kennedy's assassination. In fact, Tinker says, "I think of them like the Kennedy family."

The other new idea draws on an MTM two-hour television movie of a season or two back, "Just an Old Sweet Song," about a black family from Detroit returning to the South. MTM has prepared a script for a sequel, and there is talk of turning it into a series. "Why not do this as a black Waltons?" Tinker describes the company's thinking, "to use the shorthand of the business."

"The biggest and crying need is for material for the early evening," Tinker says. "If it isn't Fred Silverman's frothy things, it's very hard to program early evening. We're sort of typecast. Our label is comedy."

MTM is branching out beyond comedy, and Grant Tinker is thinking more about what television hasn't accomplished than what it has. "It would be good if there were a law that *one network* had to carry something *good,* not just something commercial," he says. "Television would be better if it were on only on Wednesday nights."

Over at ABC in Century City they are feeling very little pain, not from the complaints of producers, not from the Nielsen ratings, not from the bottom line. ABC swept the 1976–77 season ratings, and for the 1976 fiscal year, according to authoritative figures compiled by *Television Digest,* the network's profits rose nearly 200 percent over the previous year. If television were on only on Wednesday nights, ABC would be ready for the competition with its present Wednesday night lineup of "Eight Is Enough," "Charlie's Angels," and "Baretta."

Brandon Stoddard, ABC's vice-president for motion pictures for television and daytime development, can take a considerable share of the credit for ABC's rating success during 1976–77. The seventeen made-for-television movies the network programmed averaged an impressive thirty-nine share of the audience, exactly the same average rating as the theatrical features the network screeened. Stoddard is a youthful-looking man in his early forties, dressed in a blue V-neck pullover and tan slacks. As he talks he sometimes sounds more like another contentious producer than one of ABC Entertainment's senior executives.

"Audiences' tastes are better than most executives' tastes,"

Stoddard says. "I think things are going on now in the audience different from two or three years ago. It responds to certain feelings in the country about a return to traditional values, a family concept, importance of the tradition of the family, personal courage, versus more disaster, more violence.

"We're going in this direction because the audience really wants to see it. In the cave they were interested in people. I just hope we can carry that out. There's no reason we can't do that in any show we have on the air. Even 'Charlie's Angels.' "

Stoddard points to "Eight Is Enough" (a Lorimar production, though he doesn't mention the producer) as a possible harbinger of things to come on television. A comedy-drama about a family with eight children, based on the nonfiction book by columnist Tom Braden, "Eight Is Enough" premiered in the "third season" during the spring months of 1977 and impressed enough people to make the fall lineup.

"So far the audience response has been fantastic," Stoddard says. "The show is not great yet because it's a difficult show to do. If that show works it will have a profound effect on television. It's terribly naked all the time. You've got to be on target all the time, but the potential is enormous."

To Stoddard, television's biggest problem is getting scriptwriting that's up to the quality audiences are increasingly demanding. The writer, he says, "can't cop out, can't gimmick, can't hype, can't cheat. The writers are up against it." It's not fair to writers to expect them to come up with something polished and original in a week and a half, he admits, let alone to do it twenty-two times a year for programs like "Family" or "Starsky and Hutch."

Sympathetic criticism for the writers' difficult tasks, but not a nasty word about producers. Maybe network program executives don't speak out against producers because they think they themselves are the producers of the shows they broadcast—the architects, at least—while the independent producers are merely hired contractors. They take part, after all, in many of the basic production decisions. The President of the United States is reputed to have only one hot line; President Silverman may have a dozen, one for each of the production companies contributing programs to ABC's schedule.

9. Jay Sandrich: Directing Situation Comedy

Unlike movie directors, whose names are often household words, television directors are the unknown soldiers of prime time. Deservedly, some say.

The conventional wisdom, in fact, holds that television directors are little more than traffic managers, making sure the actors don't bump into each other or forget their lines. Television programming is a producer's medium, they say, or a writer's medium, or, more commonly now, a writer-producer's medium. Television critics, in marked contrast to their motion picture colleagues, almost never mention the director's name in their reviews. If actors perform well, the assumption seems to be that somehow they've managed to overcome the faceless system.

That's the conventional wisdom. But the occasional series of exceptional quality argues that the neglect of the television director has gone a little too far. Take, for instance, "The Mary Tyler Moore Show," which grew steadily in popularity and esteem over its seven-year run to become perhaps the most beloved series in American television history. The show's remarkable group of actors, of varying temperaments and styles, could hardly have worked together so brilliantly without a guiding directorial hand. That hand belonged to Jay Sandrich, who directed more than two-thirds of its weekly episodes over the seven years. "The Mary Tyler Moore Show" established Sandrich as the most widely respected television comedy director.

To watch Sandrich at work seemed the perfect way to test the conventional wisdom about television directors—to see, in other words, what the exceptional director contributes to the success of a television show.

I found Sandrich on the set of "Soap" at ABC Television Center in Hollywood. From the most acclaimed comedy series of the 1970s, he has moved on to become the first, and so far the only, director of the season's most controversial comedy show. There is no contradiction in this for him; everyone at "Soap" sees the show's notoriety as having been manufactured by an irresponsible, inaccurate press. And his willingness to take on "Soap" was strictly an advantage for the show's developers. "When you think of directors in television comedy," says Susan Harris, "Soap" 's writer-producer, "Jay is the best there is."

It is a Monday morning, the third day of a five-day production cycle that ends with a taping before a studio audience on Wednesday evenings. On Thursdays cast and crew receive a new script and read through it. On Fridays they begin blocking the action, and do the first complete run-through Friday afternoons. Susan Harris prepares the final draft of scripts on Friday evenings, paring down the dialogue so the action will run precisely twenty-three minutes. And by Monday, the actors are supposed to have their lines memorized.

But on this particular Monday, things are more ragged than usual. Several in the cast and crew, including Jay Sandrich, are just recovering from bouts with the flu. Others have had their weekend travel disrupted by an East Coast snowstorm. A few of the players still carry scripts as they rehearse. In any case, "Soap" may be the most complicated and ambitious situation comedy in television. The episode in preparation, the twentieth out of a twenty-five-show season, has seven sets and sixteen speaking parts, more than twice that of similar shows. "I don't really know what's going to happen with this show," Sandrich will repeat several times over the next three days. "This show is on such a fine line. This particular episode is too fancy."

The centerpiece of episode twenty is made up of two scenes in the Campbell household. The first is broad farce, the second tragicomedy—examples of the show's range of tone and style within a single episode, something which takes viewers a while to get used to. In the first scene, Jodie hides Bob in the refrigerator, and Chuck frantically tries to find him. In the second scene, Mary reveals to Burt that she has taken out papers to have him committed to a mental institution. Sandrich works with equal concentration on all the scenes—and other scenes present more prolonged difficulties—but it is these two which most clearly reveal his style as a director.

Jay Sandrich is a relaxed, low-keyed, soft-spoken man in his mid-forties. He wears V-neck sweaters, blue denim pants, and comfortable, black orthopedic shoes that are the butt of jokes from crew members during rehearsal. His dark hair is long and slightly gray; with his glasses and gray mustache he looks somewhat like a studious Groucho Marx. With more than twenty years of television work behind him, he has a very firm sense of his abilities and the qualities he brings to a show. He is worried not about the neglect of the television director but about the dangers of overemphasis, of creating a cult of the television director similar to that in film. Several times he expresses the preference that his communication with actors be privileged information, lest outsiders get the impression he is too domineering, too directive, too pedagogical, lest his actors read somewhere how they are mere putty in his hands. After all, unlike the film director, the television director has to work with his actors over and over again, even after the reviews come in.

The "Soap" cast is not putty in Jay Sandrich's hands. There is plenty of give-and-take between actors and director as the movements and readings of lines are developed. But there is also a subtle yet unmistakable shift in the way Sandrich wields his authority as the show moves toward taping. On Monday he will often put suggestions in the form of questions—"You want to think about that for a beat?" "Can I give you another way of doing it?" "What if you did it this way?" By Wednesday he is fully in charge. During a final run-through just before the dress rehearsal taping, he tells one actor about a certain line, "I want it thrown away."

"But Susan said . . ." the actor replies, referring to Susan Harris.

"No, I want it thrown away," Sandrich replies firmly. And it is, to great comic effect, in the taping.

It doesn't take long to discover how fundamentally different directing four-camera videotape television is from directing one-camera film. The film director makes all his takes of an individual scene at one time, and moves on to the next. The television director rehearses all his scenes over a five-day period, more like the director of a weekly short play. The film director can work intimately with his actors, communicating with them in whispers or behind closed doors. The television director often has to talk with his actors over an intercom when he's in the control booth and they're on the set, in front of not only the crew but sometimes of the studio audience. The film director assembles his picture retrospectively, in the cutting room. The television director assembles his show in advance, planning hundreds of camera setups, movements, and shot sequences before the taping begins.

Beginning Monday morning, after each scene is rehearsed, Sandrich goes through the script with his associate director, J. D. Lobue, laying out each shot: "A cross two-shot, cut to [camera] four, cut to [camera] two in a two-shot. . . ." Over the next three days he will be in steady communication with his cameramen, adjusting the height, angle, movement of their images as he views them on the video monitors, slightly altering his shot selection and sequence right up through the air taping Wednesday evening. "By the time of the show, I have the shots pretty much memorized," he says. "You don't mean to, but you do." During a break in the air taping, he speaks sharply to one of his editors: "You better stop recutting me. I know what I want better than you do."

If there was a "Jay Sandrich touch" on "The Mary Tyler Moore Show," it is even more pronounced on "Soap" because of production differences between the two shows. The "Mary" show was shot with three film cameras, in contrast to "Soap" 's four tape cameras. Working with three-camera film, Sandrich set up the camera placement and movements, but that preliminary work generally completed his responsibility for the look of the show. All three cameras rolled throughout the performance, and an editor put the show together from the three separate takes.

"On tape," says Sandrich, "I'm actually doing the editing. With tape you have to be more sure of what you want. You have to be more technically minded. You can get by on film without being technically minded. On tape you're the only one who understands the completed picture."

"With film," he continues, "once you do the camerawork you can pretty much forget about the cameras. But in tape you really have to learn that you have to be able to switch your focus. I'll see things I haven't seen, whereas in film your whole concentration can be on the acting. Tape is more fun for the director. There are more decisions; you're much more part of the performance; there's adrenaline flowing. You don't have the ability to concentrate as much on the acting. You have to *remind* yourself that the important thing is the acting."

Sandrich's camera style makes more demands on actors, but the actors also can display their talents better than in the more conventional ways situation comedies are shot. His cutting is much more rapid than in most other comedy shows, and he concentrates on head-and-shoulder close-ups. He calls forth more animated responsiveness from his actors at times when they don't have lines to speak, and he rewards them with more close-up intercutting. "Soap" 's expensive and varied sets hardly ever are exposed in their full glory to home viewers through establishing shots.

"I really don't believe in long shots in television," Sandrich says. "Television is more of a close-up medium than film. In a dramatic show you can do more with a camera than in a comedy. Comedy is a medium of eyes and words. As director, you make sure not to cut away from somebody too early or too late. You make sure you get the reactions of the other actors. You have to have a certain flexibility in your cutting pattern. The technical thing, once you learn it, is filled with patterns. You try not to become predictable. An audience is ahead of you. Subliminally, viewers know it's going to be a close-up now."

"One of the things I try to do is keep the performers moving," he adds. "It's easier to emphasize if they're moving. If there's movement you can vary the shots. First, stage it so the scene works. From that, it tells you how to shoot it."

The television tape director has more on his mind than the stage director, because he has camerawork as well as staging to plan; and more than the film director, because he has four cameras instead of one, and he's directing scenes instead of shots or takes. And he knows that his virtuosity, his genius, or lack thereof, is not what audiences are interested in. "When a comedy is over," Sandrich says, "you don't want to say, 'What a well-shot show.' You want good performances, a satisfied audience."

Good performances. The important thing is the acting. These words crop up over and over in Sandrich's conversations about his craft while he's preparing episode twenty of "Soap." His development of the kitchen scene reveals how he defines and works to get good performances. The scene is "Soap" at its farcical best. After Jodie puts Bob in the refrigerator, Chuck, in a state of panic, turns a grapefruit, a banana, and an English muffin into substitutes for his ventriloquist's second self. Jodie grows angrier and angrier at Chuck and seems about to do him violence when Burt enters and begins a wonderfully wacky monologue about bagels as "one of mankind's most dangerous foods." (Billy Crystal plays Jodie, Jay Johnson plays Chuck and Bob, Richard Mulligan is Burt.)

Jay Sandrich is worried about a certain tendency in actors to overplay a scene. "Soap" sometimes doesn't work for first-time viewers (myself included), because they have no way to gauge the tone. You grow into the show after seeing two or three episodes, not only by learning more about its characters but also by learning to grasp its manner, its style, the evenhanded, natural way it accommodates the weird and improbable. One of Sandrich's tasks is to maintain that naturalness in the face of tendencies, in the material as well as in the actors, to overexaggerate. There is more humor in being grotesque and thinking you're normal, and "Soap" 's triumphs come when the show maintains that tone.

At a Monday rehearsal of the kitchen scene, Sandrich tells Billy Crystal and Jay Johnson, "I know you're going to find this hard to believe, but the scene became too 'sketchy' "—meaning more like a revue sketch than the everyday household atmosphere

"Soap" strives for. (During the Monday afternoon run-through for producers Susan Harris, Tony Thomas, and Paul Junger Witt, Witt also finds the scene "too frantic, too psychotic.") At Tuesday morning's rehearsal, the cameramen are present for the first time, and the kitchen scene has them breaking up in laughter. But Sandrich suggests to Mulligan, "Rich, you know what I think we're doing? I think we're getting too cute."

Once again, during the Wednesday afternoon run-through, Sandrich says to Billy Crystal, "I'd rather underplay." Just before the first performance before an audience, the 5:30 Wednesday afternoon dress rehearsal taping, he tells the actors, "Make it as real as the conception will allow." Clearly his bets are hedged. "OK," he says, as the actors begin in the kitchen, "now we come to the Howdy Doody scene." But it draws a lot of laughs, and he says, "This is a very good audience."

Jay Sandrich was born in the "company town" where he has made his career. His father was Mark Sandrich, who directed several of the classic Fred Astaire-Ginger Rogers musicals for RKO in the 1930s, including *The Gay Divorcee* and *Top Hat.* Jay studied motion picture production at UCLA, and gained further film experience during his army stint in the Signal Corps. Then he went into television in the fifties as an assistant director on "I Love Lucy" and has worked in comedy ever since. He produced the first year of "Get Smart," among other credits, and got his start directing on a show called "He and She." It was his work there that brought him to the attention of Grant Tinker at MTM Enterprises when "The Mary Tyler Moore Show" was forming. "Originally, I told Grant I'd do the first four or five," he says, "but once I started doing 'Mary' there was no other."

Why aren't there more directors like him in television, experienced, skilled, and respected (at least within the industry)? "Eventually directors want to become producers," Sandrich says. That's where the prestige and power in television are. "I started as a producer, so I'm very satisfied. I love what I'm doing. I really know this is where I want to be. I don't feel frustration in being a television director.

"The writer-producer's job is the hardest in television," he continues. "A lot of tape work is done on Saturday and Sunday. You always have a problem as a producer. As a director, my problems exist at a particular moment, and I deal with them. As a producer, I'd wake up at four in the morning and realize I had problems. Plus, I enjoy working with actors. I find enough creativity with that."

The director brings to a television show what producers, writers, and actors can't, a sense of staging, of movement, of ensemble performance. "My joy comes in taking a scene and making it come to life," Sandrich says, "creating an atmosphere that actors can thrive in. The director's function really has to do with storytelling. Sometimes you can come up with an approach, a point of view, a scene. A director can't just do what's handed to him.

"A director has to try and help the writer," he explains. "You have to focus on what the problem is. The director can do that because he does the scene over and over. You have to prod the writers into going back. I've been fortunate. I've worked with the best comedy writers in television, writers concerned with human emotions. Sometimes, however, brilliantly written scenes do not work as well as you think."

Still, it seems clear that very few television directors bring the vision or authority to their work that Jay Sandrich does, and when they do, viewers and critics are not attuned to understand the director's contribution. One of the problems with television, as Sandrich sees it, is the sheer volume of production. There are so many shows, so much demand for material and directors, that a lot of work is inevitably, as he politely puts it, "less than brilliant." Another problem is that most television directors lack the opportunity he has had to develop and to stay with an individual show. "Normally most shows have various directors to do episodes," Sandrich says. "There's no stamp on a show, no style."

Yet television may never become a director's medium in the same way feature films are now held to be. "In features," Sandrich says, "if a director doesn't know what he's doing, the movie will fail. In television, once a show is cast, the script is the most important thing. If you have a wonderful script, you will have a good show even if it's poorly directed." (But with poor direction, he points out, the actors may have trouble, and it's hard for the editors to put it together.) "With a good director, it may have been a wonderful half hour."

And why is the director's shaping hand so invisible to those who write about television? "People don't tend to think of television as an art medium," Sandrich says. "I have personally spotted directors, who have become top feature directors today, the first time I saw their work on television. I look for: Are the performances, overall, moving? Does it move? Is it crisp? Is the camera in the right place? If you can answer these questions, it's good direction. Once in a while you'll see things that only a director could have done."

Over the last two days of the "Soap" production cycle, Tuesday and Wednesday, Sandrich spends most of his time in the control booth, periodically returning to the set to go over points with the actors face to face. He sits at a control panel with an array of microphones, dials, and buttons, and faces a wall of monitors—four black-and-white images for each of the cameras. Above them, in color, is the master image, the shot home viewers will be seeing. At his left are Linda Day, the script supervisor, who times the scenes and notes Sandrich's comments about acting or stage problems, and J. D. Lobue, the associate director, who supervises the cameramen and stage crew. At his right is Gerry Bucci, the technical director, who presses the buttons changing camera shots each time Sandrich gives the signal. During run-throughs his signal is a small toy metal cricket. In the two actual taping performances, he eschews the mechanical aid and snaps his fingers for shot changes, more than six hundred times in all.

As they move closer to the 5:30 dress taping, an aura develops in the control booth not so much of tension as of firmer command. Detailed instructions are pouring out from Sandrich to the actors and to the cameramen: "Jennifer, finish your line before you look down . . . tighter, D. J. [on camera two] . . . little higher, Jan [camera four] . . . D. J., you're favoring her, make it fifty-fifty." Someone produces a cassette-sized computer football game, and suddenly the director, associate director, and technical director are busily pressing buttons on the toy in their spare moments—a busman's holiday with instant gratification. At one point during the dress taping while cameras are being repositioned between scenes, Sandrich racks up a big play. When they're ready to roll tape again, he exclaims in mock annoyance, "Sure, start the tape when I'm in the middle of a long run." At the next scene change, he scores.

Taping television comedy before an audience is another mark of difference between television and film, and even between television comedy today and television comedy in the days of "I Love Lucy." "Audiences bring an unknown element," Sandrich says. "An audience forces actors into things that wouldn't happen on a silent sound stage. When you're in front of an audience, you can't believe you can count on the laugh track. You don't settle when you have to face people."

Some filmed television comedies were performed before an audience, but the shift to four-camera videotape production made possible the innovation of the dress rehearsal taping. "It's a wonderful thing to be able to do two shows," Sandrich says. "It's the one big advantage tape has over film. When you do a film show, you have to take what you get. Film is too expensive to do two shows. From the two shows, you can make a third show that's better."

He is referring to the common practice of selecting the better of two takes for the final version of the show. On "Soap" this is further refined by keeping all four cameras on ISO (for "isolation")—staying on the shot even when it's not on the air—and taping the complete run. This sometimes produces an unplanned reaction shot or serendipitous bit of business that can go into the final tape edited for broadcast.

The 5:30 dress rehearsal audience, for unexplained sociological reasons, is generally considered to be less sophisticated than the 8 p.m. air taping audience, more likely to laugh at broad, corny humor. On this occasion, however, everyone seems surprisingly pleased with the audience response, until the Campbell living room scene. Through rehearsals Sandrich had been increasingly stressing the poignancy of the scene, in which Mary reveals she has decided to have Burt committed, and Burt comes to believe he is mentally ill. Just before the dress taping, Sandrich had remarked how sad the scene played, and said, "We're not going for laughs."

But the audience doesn't read the scene that way, and howls when Mary comes down affirmatively on the word "commit." "What are they laughing at?" Sandrich exclaims. "They really felt we were doing a comedy scene."

That dissonance aside, the dress taping goes extremely well, and the cast retires to a rehearsal room where a buffet dinner awaits them. While the players are eating, Sandrich and the producers, sitting at the head of a horseshoe-shaped table, speak with the cast about problems they had noted. It's instructive of producer-director relations on this particular show that Sandrich is clearly in charge of the session. They are basically satisfied, however, and there are no rough moments. To Richard Mulligan and Cathryn Damon, who play Burt and Mary in the living room, Sandrich says, "I thought the scene was lovely." To all he says, "Good show, good dress."

The air taping, however, does not start so auspiciously. In the second scene Sandrich has to break in on the intercom and say, "I'm sorry, we have to stop. We have a slight technical problem. We blew our lines." The audience lets out an "ooh" of disappointment. They begin a pickup take, but that goes unsatisfactorily, too, and Sandrich finally has to go out to the set to instruct the actors. In the kitchen scene, Jay Johnson, while turning a grapefruit into a ventriloquist's dummy, misses a line, and they stop again. Sandrich calls out on the intercom, "It was the grapefruit's fault, Jay." Johnson replies, "It's not union."

There are further problems—missed cues, overacting by an extra, unheard lines—in almost every scene. Rather than reshooting in front of the audience, Sandrich decides to hold off some of the retakes until after the audience has left. In the end, as the last pickup is shot, he joins in laughing at the successful comic performances of his players.

Jay Sandrich has done all he can to bring movement, pace, and subtlety to episode twenty, to make the director's contribution. The next morning will bring a new script and new problems. What was done or undone on the completed show will not burden him. "This is not a matter of life and death," he says. "This is a television show. This is entertainment. We all have other lives. We all know this is not going to affect the world. You might as well enjoy it."

10. Blacks in Television, Blacks on Television

Nearly two years have passed since the "Roots" phenomenon and the prediction, fostered by the most watched series in television's history, that the image of blacks on prime-time commercial television would henceforward be more realistic, more honest, more rooted, pardon the pun, in the realities of everyday black existence. The new television season of 1978–79 is now at hand to mark, given the pace of program development, the first test of that prediction. And what can be found on the networks' fall schedules? More "Dy-no-mite!" and rotund Rerun and gabbling George Jefferson.

Times change, and so does our perception of television. Earlier in the seventies the producers of shows like "Good Times," "What's Happening!!" and "The Jeffersons" were lauded for giving us, within the familiar conventions of the situation comedies, blacks who experienced life the same as everyone else—they could have feelings of love or anger, bicker with their children, be petty or small without turning into buffoons like Amos 'n' Andy. Above all, they could admit that race was an issue in their lives. From today's perspective, those same characters are looking more and more like Stepin Fetchit and Jack Benny's Rochester. The trouble is the characters haven't changed, and we have.

During the 1977–78 television season, though prime-time sexual titillation occupied the attention of media pundits, reform groups, and the practitioners of television sociology, the impact of "Roots" was also creating a new undercurrent of concern about television's presentation of blacks. Did Kunta Kinte suffer his ordeals in order to bring forth "Baby, I'm Back"? "Baby, I'm Back" will not be back, but it was, according to the Nielsen ratings, the most popular of all the black series.

The "Roots" phenomenon remains, for the time being, just that, a phenomenon. (The repeat of "Roots" and "Roots: The Next Generations" don't change that.) Over the past year or so, two hour-long dramatic series featuring contemporary black characters were developed by two of the most important production companies in Hollywood for consideration by network programmers. They differed from the existing sit-coms not only in theme and treatment but also in the fact that significant behind-the-camera personnel, as well as performers, were black. Neither made it onto the regular prime-time schedule.

One, "Kinfolks," an MTM Enterprises project, never even got an airing on CBS. The other, "Harris and Company," produced by Universal Television, was screened by NBC as a two-hour pilot (called "Love Is Not Enough") on "NBC Monday Night at the Movies," and the network, though turning it down as a regular series, did order four more one-hour episodes. These will be ready on the shelf for the inevitable failure of some other show. Four one-hour episodes of "Harris and Company" are hardly enough effort by the networks to satisfy either the expectations created by "Roots" or the changing values and attitudes of the television audience. No one is quicker to make that clear than the black men and women who are working on black shows or elsewhere in television.

Stanley Robertson, the executive producer on "Harris and Company," is one of the few blacks in a top television management position. He joined Universal as a producer two years ago after spending nearly twenty years at NBC, rising from a page at the Burbank television studios to a network vice-president in charge of motion pictures for television. He is an athletic-looking, outspoken man in his mid-forties and dresses, as do all Universal executives, impeccably in suit and tie. I talked to him when the first rough cut of the two-hour pilot show was ready for screening.

" 'Love Is Not Enough' is the first black two-hour pilot ever made for a network," Robertson says. "Frank Price [then the president of Universal Television] put up $1.3 million of Universal money to do the pilot. That's a lot of corporate money to put up." Robertson developed the concept along with Arthur Ross, a white writer, who did the script. Ivan Dixon, a black, was hired to direct.

Robertson's attitude is a refreshing combination of optimism and candor: Optimism you can always find in television executives, candor is more rare. He was a lowly page at NBC back in the late fifties when that network canceled "The Nat 'King' Cole Show" because sponsors would not put their ads on it. "I have seen change. I have seen improvement. Things are getting better." But, he goes on, "there's still not enough. Let's be honest. There are still pockets of resistance. There are still people in the business who are not committed to equality. It would be easy to single out the entertainment business. There are people who worked on the pilot who didn't like the idea of a black producer and black director.

"There are still people in the business who have these old attitudes," Robertson says. "There's more prejudice than meets the eye. It's not the presidents and vice-presidents, it's the guy at level eight, nine, or ten—the guy who hires the driver or the grip. Those kinds of things affect the product. The head of the company is committed to depicting society as it is. When it gets down to an assistant producer in the casting department, he won't do it. It's in the lower echelon."

The "Harris and Company" series is not about the presidents and vice-presidents, it's about a guy in the lower echelons. Mike Harris is a black blue-collar worker at an auto supply company in Detroit. (Bernie Casey, the onetime pro football star, plays Harris.) His wife has died and left him to care for five kids ranging from preschool to late teens—he's the Harris, they're the "Company." Encouraged by a cousin, Harris decides to pull up stakes and move to Los Angeles. Once there, the job his cousin promised doesn't materialize, but he hooks on as a tow truck driver for a white couple who run a gas station, fixes up a run-down tract house, and begins a new life.

"These are people like everybody else except for the ethnicity," Robertson says—ethnicity apparently being a television code word for race. "They're blue-collar people trying to make it in twentieth-century America. It's meant to make you feel good. I did not want to do an all-black show. I did not want to train an all-black crew. I want that home flavor to permeate the show. We don't live in an all-black society, an all-white society. I'm a strong believer in the fact that we live in an integrated society."

Stanley Robertson is a man who takes pains to get things right. In the pilot, when Liz, the eldest Harris child, goes to inquire at a San Fernando Valley college, the admissions interviewer is an Oriental—a small point, but still an effort to portray Southern California's ethnic diversity. (During our conversation he mentioned another show, set in San Francisco, where no Chinese ever appear.) At times the integration theme may be carried too far in the pilot—anyone who knows contemporary Detroit will wonder where those white neighbors came from who buy furniture at the Harrises' going-away sale on the lawn of their housing project.

What made the pilot special, however, was its economic honesty. These *are* people like the rest of us, who worry about budgets, who don't have enough money, who can't afford their favorite foods. "We're not gonna stop loving each other," Mike Harris says to his children after a small family crisis. Harris and Company have plenty of love for each other, but love is not enough: They need money.

"There are too many black comedies on," Stanley Robertson believes. "The spectrum is not balanced. Because of the preponderance of comedy, the American people have got the idea that black people are funny. Not that people have to be serious and down at the mouth. I've had people say, 'You can't do a series, a black series.' People want to be emotionally involved. Maybe, except for 'Roots,' we haven't had the opportunity to see blacks get emotionally involved. If you did 'The Jeffersons' dramatically, it might work, if you could accept those characters, if you got involved with them."

Previous attempts at dramatic shows about blacks have failed, Robertson says, because the leading actor was not strong enough. In "Harris and Company," he believes he has both the concept and, in Bernie Casey, the performer. "My feeling is it does work, we do have the right actor." Television audiences may even get a chance to see for themselves if he is right.

No one got a chance to see "Kinfolks." The pilot episode was produced for the 1977–78 season, and after waiting in vain to see if CBS would give it a try as a replacement series, MTM released the cast. In contrast to the Universal pilot, "Kinfolks" had a white producer, Philip Barry, and a white director, Fielder Cook, but a black writer, Melvin Van Peebles. And in contrast to Stanley Robertson, who has spent his entire career in television, Melvin Van Peebles is a man whose appearance on television, even as a writer, is something of an oddity. For as Van Peebles himself says, "I'm banned from television, the talk shows and so on. I thought it would be interesting to get on television the only way I could, by writing for it."

If Van Peebles is indeed banned from the talk shows, it is because of his reputation as a militant black playwright and as the producer, director, writer, and star of the controversial 1971 film about an urban black rebel, *Sweet Sweetback's Baadasssss Song*. I talked with him in the offices of his production company, Yeah, Inc., on Seventh Avenue in New York. A lean, lithe man, dressed in a purple shirt and black trousers with red suspenders, Japanese sandals on his bare feet, Van Peebles came to television with a specific story to sell.

"I got a letter inviting me to a cookout down South," Van Peebles says. The cookout was for a reunion of a Van Peebles family, and Van Peebles suspected the family was white and had made a mistake, though he never went down to find out. He did, however, develop a story idea from the incident and took it to an agent. Eventually the incident made its way into the "Kinfolks" pilot, where a group of white Simmonses shows up at the black Simmons home for a "family reunion." The white adults are befuddled or indignant, but a little white girl hugs the elderly black great-aunt, and a point about racism is made.

The agent soon reported back that a producer was interested in Van Peebles, but for a different idea. The producer was Philip Barry of MTM Enterprises, and the result was "Just an Old Sweet Song," an award-winning television drama written by Van Peebles and directed by Robert Ellis Miller. Cicely Tyson, Robert Hooks, and Beah Richards starred in the story of a black family from Detroit that visits the wife's dying mother in the South and decides to stay.

"This New South probably ain't what it's cracked up to be," the husband says in the show's coda, "but it's better than it used to be. Everything we wanted up there is down here. I still hate the South. The South is up North in them ghettos, but the program is still the same—ripping off the black man."

(It's instructive, by the way, to compare the brief images of Detroit we get at the beginning of the "Harris and Company" pilot, all gleaming new Renaissance Center and panoramic long shots, with the dirty, crowded, run-down Detroit black ghetto of "Just an Old Sweet Song.")

The idea for "Kinfolks" grew out of the success of "Just an Old Sweet Song." In the pilot show for "Kinfolks," Robert Hooks continues as the husband, Madge Sinclair replaces Cicely Tyson as his wife, and Beah Richards, formerly the dying grandmother of "Just an Old Sweet Song," is reincarnated as the great-aunt. Van Peebles got a chance to include his original television idea, the episode of the family reunion. But the story line turns on the murder of Nate Simmons's white business partner by a demented red-neck and the threat by the murderer's brothers to do further harm to blacks—stereotyped bigots and conventionalized violence, as I viewed it.

Van Peebles gets annoyed at my suggestion that he may have had to adapt his views to the straitjacket of formula television. "If you look at my preachments over the years," he says, "they've been exactly the same. *Sweetback* was as American as apple pie. There was an audience out there, as the money I made showed. You don't call Alice Cooper and say, 'So you've given up snakes.' Why must I be limited? There is a continuing line in all my work—which is cool."

Nevertheless, Van Peebles makes clear he sees television differently from other forms in which he works. "I work within the framework of television. If you think you're going to write 'The Exposé,' you're pissing up a wet road. I didn't approach television with that. I never had any illusions about television when I went into it. I have the option to do my deepest, strongest work elsewhere, so I don't have to do it on television." He expresses good feelings about his work on "Kinfolks." "Many of the hassles are not indigenous to blacks," he says, "they're indigenous to television per se."

Beah Richards was one of the black performers in "Kinfolks" who knows how television narrows one's options and also how to create options for herself outside television. A poet as well as an actress, she won an Emmy for her production "A Black Woman Speaks," a performance piece based on her own poetry which she has presented both on stage and on public television. When I went to talk with her in her Los Angeles home, I expected to find a woman a good twenty years older than she actually appears, because I had just seen her as the elderly grandmother in "Just an Old Sweet Song" and the old, blind great-aunt in "Kinfolks." Instead of the decrepit old folks of her television roles, I found a vigorous, eloquent, humorous woman who lives surrounded by plants, dogs, and African art.

She had also just completed, as it turned out, a role in a television film, "Ring of Passion," about the Joe Louis-Max Schmeling boxing rivalry of the thirties, produced by Twentieth Century-Fox for NBC. She played Joe Louis's mother. "I've been everybody's mother," Beah Richards says. "It always happens, doesn't it? You get a rash of them." And she vigorously scratches herself.

Beah Richards has been striving for years to get rid of the rash of stereotyped roles for black performers in movies and television. "I don't think I have seen much of a change," she says, "but I have seen some." She first came to Hollywood in 1964 to perform on stage in James Baldwin's play *Amen Corner,* which lasted a year and a half. "The industry discovered I was here and offered me one day's work as a maid in a picture. Attitudes change slowly, but they do change if we insist upon it. You face the business of knocking on the door when you have the possibility of getting a script you don't want to do. And yet the doors must be opened."

Going from roles as maids to roles as mothers is how Beah Richards measures how far she has opened the door. She won an Academy Award nomination for best supporting actress for her performance in *Guess Who's Coming to Dinner,* where she played, yes, Sidney Poitier's mother. With her long list of motion picture and television credits, she has concern for the young actors for whom the doors are just opening now.

"The actor is a kind of sitting duck, you know," she says. "Many black people are going before the camera for the first time in their lives, you know. They have one day to prepare. There are no schools to prepare you. We have training schools, but they're no substitute for the real thing. The crews are sitting around while the actor deals with his nerves."

"Kinfolks" pleased her. " 'Kinfolks' is a gem of a piece about people who are flying from the violence to find a little peace," she remarks. Actually, they seem to be flying from urban violence to rural violence in the "Kinfolks" pilot, but Beah Richards is referring to matters of ambience rather than of plot. Born in the South, she is thinking of the show's contrast of the tempo of life between urban Detroit and the rural South.

"The pace in the South is different from the East and the West," she says. "The pace in the South is much slower. Thinking before you speak, weighing—these were taught us by years and years of compromise. You dare not take the time that you know is there in that moment." She recalls, as a young girl, addressing her father. " 'Daddy.' Silence. 'Daddy.' 'I heard you the first time.' There's a little of that in 'Kinfolks' and 'Just an Old Sweet Song.' It's there in the script."

Did it get lost in the production? When I asked Melvin Van Peebles about certain elements of tone in "Kinfolks," he replied shortly, "I didn't direct it." But Beah Richards speaks favorably of Fielder Cook, the show's white director, as a southerner. "The lines of communication are not as alien. We know each other in the secret places. Won't it be wonderful when we can begin talking about that?"

"I think for all of us it is a time of tremendous opportunity,"

Beah Richards says. "The ground that has been won in opening the doors in Hollywood, it's an intellectual revolution. We're going to have to redefine concepts. Black is not a color—nor is white. We're going to have to think of black as a point of view and see that that point of view is heard. The artists who express that point of view are the ones I want to work with—and be free."

As blacks move into the mainstream of television production, does a black point of view show up, too? The answer, of course, is mixed. Universal Television is one of the production centers that has been moving blacks rapidly into positions of authority. Stanley Robertson came over to Universal from a senior position at NBC, but two other young blacks worked their way up at Universal: Charles F. Johnson from mail clerk to coproducer of "The Rockford Files," the only black producer on a prime-time weekly series during the 1977–78 season; and Yvonne Demery from secretary to associate producer on "What Really Happened to the Class of '65?" Both are articulate, modest but self-assured. They want to be known for their abilities rather than for their color, though, as Johnson says, "You don't ever forget the fact of color."

Johnson and Demery are close friends and collaborators, but it would be a mistake to lump them together as harbingers of a new black generation in television, simply because of another fact, the fact of gender. Demery's progress as a woman in television production is almost as important as her progress as a black. And her career probably received more impetus from the women's movement—and from federal affirmative action programs—than from any organized pressure on the studios to promote blacks.

"The secretaries are erecting a statue to me in the lobby," Demery jokes. "I'm the first secretary in the company to move up." She admits there is sometimes a conflict between her allegiances as a woman and as a black. "The black women are very happy" about her achievements, she says. "The black men . . . eh?" A shrug.

Raised in Pelham, New York, Demery studied at the University of Bridgeport, did social service work in New Rochelle, and joined a singing group, the Chantels, that toured American military posts in Asia for the USO. The National Urban League helped get her a secretary's job at Universal in 1967, but after several years that palled and she went to work for an independent film production company and a television station to gain production experience. When Universal was looking for women to place in a virtually all-male production department, she returned to the company as a production coordinator.

Johnson also got his start at Universal in production as a coordinator. Another easterner, from Middletown, Delaware, he was sent at age fourteen to a private boarding school, as the second black to attend. He earned bachelor's and law degrees from Howard University, though he always wanted to be in the entertainment field. Armed with an article in which Steven Spielberg told how to get onto a studio lot, he made his way onto the Universal lot. "I walked all around the lot," he recalls, "ate in the commissary. Hal Holbrook stopped and talked to me." Soon he was working in the mail room, but that only lasted until the studio discovered his educational background—all of four days. He moved up.

"It's a great education being a production coordinator," Johnson says. "You're the liaison between the Tower management [in Universal's headquarters building] and the production unit. You're clearing scripts, making deals with writers, going to dailies every day. You learn what you can and cannot do physically with film. You watch and you ask how things are done." When "The Rockford Files" was begun, he became production coordinator for the series, then associate producer, then coproducer. "I never wanted to be anonymous from the black aspect," Johnson says circumspectly. "I must point out I didn't know how to do that." In other words, he had experience from prep school days of being the only black in a formerly all-white situation.

Neither Demery nor Johnson has yet had experience on a black television show, but they both want it. Together they have worked on a program concept about a black attorney. "I have a great desire one day to do a black dramatic series," Johnson says. "Yvonne and I wrote a pilot treatment and have worked with the Tower management on that. The lead character has parity in all fields, all walks of life."

"We still have designs on getting it done," Demery says. 'It was a show that we wrote with Louis Gossett in mind, long before 'Roots.' "

The "Roots" phenomenon is fading into history, and black blue-collar workers are having a hard time getting on the small screen, let alone black attorneys. "The comedies served a purpose," Johnson says, "but the tokenism of the sixties is passing. Television to me is behind the times. There are so many prototypes in society who can serve as models for television."

Demery concurs. "I'm not that happy with the black image on television. The image was successful as long as blacks were being laughed at in comedy. We felt that the public was ready to accept more. You could extend the public's imagination to accept the fact that I'm a black producer. Our life-style is not stereotypical of anything."

Television's black producers have extended their imaginations, we've extended our imaginations. Maybe it's time for the people in network programming to do a little stretching of their own.

11. The Liberal Education of Dick Cavett

"Entertainmentt," read the subway posters. At every station stop on my way to "The Dick Cavett Show," that word appeared above Cavett's smiling, larger-than-life image. After a few days' visit to public television's weeknight celebrity talk show, I decided that the copywriter who spelled entertainment with a double *t* was more than merely clever. He had recognized what everyone connected with the show, including Cavett himself, readily admits and even takes pride in: the show's double character, sometimes serious, often not.

That quality of doubleness extends, of course, to Dick Cavett himself. "His eyes are very blue. I never realized they were that blue," a woman sitting next to me in "The Dick Cavett Show" audience says to her companion. And in the next breath she adds, "He's so intellectual, so well read." Cavett is a good-looking man, a performer, a celebrity in his own right. He also bears an intellectual mien, an aura of being conversant with books, theater, music. The several sides of Cavett's personality no doubt give the show its variable and still undefined nature.

But what makes a human being interesting doesn't always make a television show work. As "The Dick Cavett Show" enters its second year in fall 1978 on PBS, its special word is no so much "entertainmentt" as it is "survival." Perhaps that's as much as one has any right to expect from a show that marked something of a new departure both for Cavett and for public television. The new departure, to mix a metaphor, had a hard time getting off the ground.

After opening in 1977 with a rather crude grasp for notoriety (centering on whether Italian actor Marcello Mastroianni spoke the word *worker* or an obscenity), the show stumbled, lost momentum and viewers, changed directors, and slowly began to find itself late in the spring, just before stopping production for the summer. Its most heartening news was the strong support it received from public television stations, whose managers voted to increase by 50 percent their contribution to its budget for the second year. Now it remains to be seen whether "The Dick Cavett Show," a talk show with a difference, can make "entertainmentt" a synonym not for survival but for success.

Dick Cavett's rendezvous with public television was not exactly a destiny he warmly embraced. "I don't mean to make this sound too grim and negative," says Christopher Porterfield, Cavett's college roommate, longtime friend, and producer of "The Dick Cavett Show." "But by a process of elimination and because it was the next logical step," Cavett offered his services to PBS. Porterfield adds that Cavett at first resisted "because it might look like a step down."

The previous couple of years for Cavett had been, as Porterfield says, "not a productive time." After ABC dropped its edition of Cavett's late-night talk show, Cavett was quickly picked up by CBS. The person responsible for bringing him to CBS was Fred Silverman.

"Silverman had the idea that Cavett had an easy likability that would go over in prime time, like Arthur Godfrey," says Porterfield. "Somebody who didn't do anything but was surrounded by people." But when Silverman soon moved to ABC, his successors "regarded themselves as saddled with a holdover contract," according to Porterfield. Over the eighteen months of Cavett's tenure at CBS during 1975 and 1976, he was in one prime-time special, four prime-time variety shows, and made several guest appearances. The variety shows were a summer replacement series, and they bombed.

"It was a disaster," Porterfield says, "because it was in the hands of people who had no idea what kind of animal Cavett was." It was as inimical an atmosphere as could be for Cavett." And what kind of animal *is* Cavett? Porterfield: "He can tell a few jokes, but he's no threat to Bob Hope. He can be a host, a catalyst, and a front for other things to happen."

Thwarted from becoming this generation's Arthur Godfrey, Cavett fell back on what he had already done. "He wanted to have a talk show again," Porterfield says. "There was a long flirtation with a syndicator. In the end that came to nothing, because they really weren't able to trust talk. Cavett's whole career is built on the notion that talk *is* entertainment."

That's when PBS suddenly seemed palatable. From Chloe Aaron, vice-president for programming, Porterfield says, "we learned the facts of life. You can't have a show out of wedlock. You have to marry a station first." In other words, only individual stations, not the PBS network, can offer programs to other public television stations. Cavett's company, Daphne Productions, tied the knot with WNET, New York, in a shotgun arrangement that, as Porterfield concedes, has a lot of "wariness and prickliness on both sides."

Daphne retained "what is grandiosely known as artistic control"; WNET furnished the studio and agreed to pay the "below-the-line" production costs. PBS stations put up a million dollars, Gulf & Western Industries and the Chubb Group of Insurance Companies each contributed a quarter million, and the man the Germans call "The King of the Talkmasters" was back in his brown leather armchair, talking.

The ideal talk show, to hear Cavett's staff talk about it, resembles a living room conversation. For Cavett, the ideal talk show situation may be more like a classroom. The one persona Cavett seems to wear most comfortably is the ever youthful Yale collegian, the handsome, talented, midwestern boy who stepped out of an F. Scott Fitzgerald novel into the Ivy League and the smart set of New York. Or maybe I got that impression only because I watched Cavett do two shows with Paul Weiss, who was once Cavett's philosophy professor at Yale.

Cavett was bringing Weiss back for a second set of shows, in part because Weiss's first appearance—in which the philosopher had opined, among other things, that there were no great women philosophers—had generated one of the largest volumes of mail of any guest all year. But Cavett also seemed to slip into a subordinate role with Weiss, different from his manner with other guests. He became the diffident and questing student, trying eternally to earn praise from the professor who is also a friend. And Weiss, a short, bald, energetic man with a quick-witted, feisty style, possessed the professorial wiles to jar his old student and friend into discomfort—and to provide a rare glimpse of the talkmaster at a loss for words.

Along about the middle of the second show—Cavett tapes half-hour shows with the same guest one after the other, to be shown on consecutive nights—Weiss leaned forward and said, "Dick, can I ask *you* something?" He observed that Cavett, during years of talk show hosting, has conversed with artists, writers, academics, politicians, musicians, men and women of wide accomplishment. "What have you learned," Weiss asked, "from your guests?"

Cavett suddenly looked like an unprepared student who didn't expect to be called on. "How much time do we have?" he asked with a forced jocularity. Then, "I can't answer that." And after another pause, finally, "I almost see myself as an actor."

Weiss, swift and implacable, leaning toward his pupil, jaw jutting, asked, "Today, too?"

"Who was on?" Cavett mused, addressing himself. "And you can't think of any of them." Then he said, "There must be an easier way to make a living. I wonder why I don't more often learn something."

A little softer now, having broken through Cavett's veneer, Weiss asked, "Do you feel that you have gained in knowledge? That you are more stable?" Was this some reference to Cavett's collegiate past? No matter, the water was getting too deep, and Cavett steered the dialogue to a safer spot. At the show's end he asked Weiss, "Can I have a makeup exam on the question you asked me? I didn't do very well on that."

Several days later I asked Cavett what answer he would give, with time for reflection, to Weiss's question. "I don't think there is one," he said. "It is a liberal education in some sense. It's a little like school. You have to study some things you otherwise wouldn't. It's like a continuing education." Then he added that he often feels like a student, wanting to read "anything but the assignment" the staff has given him to prepare for interviewing a guest.

"The Dick Cavett Show" is taped in WNET's Studio 55, on Ninth Avenue and Fifty-fifth Street in New York, a few blocks from the offices of Daphne Productions. Nobody on Cavett's staff is happy with the setup, which comes as part of the deal with WNET. There's no backstage, and Cavett can't make an on-camera entrance.

Nor is the studio really built for an audience. In fact, the show began in 1977 without a live audience, and that was later perceived as one source of its problems. Temporary stands and folding chairs that seat about a hundred were put up for each Cavett taping, and an audience invited in. By the end of the season, the stands were regularly filled.

The idea to go on without an audience was Cavett's. "The few times I had done a show with no audience, I felt relieved, more relaxed," he explains. "It did produce some very good shows, but then I just swung in the other direction. Maybe I need the audience. As an erstwhile comedian, I may occasionally go for a laugh."

"The Dick Cavett Show" airs five nights a week, but shows are taped in the early afternoon three days a week, Monday through Wednesday. Cavett may do from one to four shows an afternoon, depending on circumstances. In the case of one guest, William Safire, the *New York Times* columnist, a decision was made immediately after one show to tape a second. As the second show began, Cavett made a remark about "last night's" show and added, "and we're wearing the identical clothes."

During the week I visited, Cavett made eight shows with five guests: Safire and Elia Kazan, the film director, two shows each on Monday; Weiss for two shows, and Maurice Sendak, author-illustrator of children's books, for one on Tuesday; and a single show with novelist Isaac Bashevis Singer on Wednesday. James Baldwin was also scheduled for Wednesday, but the staff canceled him at the last minute. Baldwin had just returned from Paris, and there was "not enough to go on," Porterfield explained.

On Monday Cavett arrives less than ten minutes before the

Safire taping is scheduled to begin. He's always late on Monday, a staff member says, coming in to New York from his weekend place at Montauk on Long Island. And he looks fresh from the beach, wearing a blue sailor hat, bright red nylon jacket, and sunglasses. He disappears into a dressing room and reappears, dressed and color-coordinated for the show—brown sports jacket, tan slacks, brown shoes, yellow button-down shirt, brown tie with yellow dots.

During the first taping, while Cavett and Safire are covering such subjects as Bert Lance, Richard Nixon, and Watergate, Porterfield writes something on a notepad, and the cue man copies it on a large cue card. Porterfield holds it up for Cavett to see: "It's quiet but *interesting* . . . probably would hold for two if you feel it . . . especially with more follow-ups." Later Porterfield writes another note: "I'm still for two." When the first half hour is over, Cavett, Porterfield, and Tom O'Malley, the segment producer, briefly confer, and the second show is agreed upon.

Among the items Cavett decides to follow up on is Safire's penchant for punning. During the first show, Safire had described a remark as a "schadenfreudian slip," and though Cavett had let it pass at the time, he calls attention to it in the second show. There follows a brief discussion of puns, which, as it turns out, is Cavett's way of putting Safire on notice that talk show hosts as well as guests can play that game. Safire, author of a first novel, *Full Disclosure,* about political chicanery in Washington, confides that he took the 19th-century English novelist Anthony Trollope as a model. "You've confirmed," Cavett says at the show's close, "that there's a Trollope behind every great writer."

Elia Kazan gives Cavett's wit less opportunity. The director and author, a practiced showman, talks to the camera instead of to his host, and pointedly refuses to tell some of his best stories, leaving Cavett frustrated, scratching his head.

Cavett's literary guests seem less concerned with which camera is on, more interested in making contact with the host himself. They aren't quite as forward as Weiss, but they're not afraid to talk back, either.

To Isaac Bashevis Singer, an elfin man with a heavy Yiddish accent, Cavett admits he was not prepared for the passion and sensuality of the elderly author's stories and novels. His impression, he says, was that Singer was prim.

"What does prim mean?" Singer shoots back—another surprise exam.

Cavett begins to say prim means straitlaced, then, wary, says, "You're not going to back me into *that* corner."

"I don't want to back you into any corner," Singer softly answers.

Was it Cavett's idea to have Singer as a guest? The answer is no. "There's a good show coming out of Singer," Cavett tells me, "but I probably would not have put him on my list." In fact, Cavett does not take part in the process of booking guests until the very end. "Sometimes," he says, "there's somebody I've just had it with. There's a streak of that in me that makes it very hard to book."

The Daphne staff meets on Wednesday afternoons, after the week's last taping, to discuss future guests. "Cavett does not sit in," Porterfield says. "He tends to be a distracting and disturbing influence. Anything methodical is alien to him. He goes off in all directions. I say that in the friendliest spirit, of course. I run the list down to him, and he exercises his veto."

Guests who are AFTRA members receive scale wages for being on the show: $205 for a half-hour program, $251 if they perform. Others get an honorarium of $100 or designate a charity to receive the payment. With such a modest stipend, the show seeks to compensate its guests by providing first-class travel and hotel accommodations and a limousine to bring them to and from the taping.

Why couldn't "The Dick Cavett Show" provide a first-class program for the public television audience when it began? "It was bewildering to us," Porterfield admits. "We did know the show had an uncertainty of touch. Dick as a performer didn't know what it was about. Nobody could have given a more devastating criticism of the show than we could. The only thing to do was make it stronger and better by our own lights."

"I knew something was wrong in the first month or so," Cavett later corroborates. "In a way I was going in discouraged. With thirty, really twenty-eight, minutes, I was trying to do the show with some of my old instincts from ninety minutes."

During that early difficult period, many station managers from local PBS stations called up to offer suggestions. After all, Porterfield explains, they were paying for the show. "There's a much more direct relationship than at ABC." But the advice was often contradictory. Some wanted the shows to tilt more toward entertainment, others wanted them more serious. Audience ratings were promising at the beginning but rapidly fell, reaching bottom in December. Then, with a new director and a live audience, things slowly began to pick up.

In return for the stations' increased contribution, Porterfield would give them a new set, a new studio, a new opening, and new music—"more pace, more energy, more aliveness in the feeling of it." He wants not to be "afraid of a bigger, theatrical feeling." He expects to break up the pattern of bookings, have more guests who are politicians and sports figures, though in general "the kind of people we've had we'll have next year." In short, there will be some tinkering.

But Porterfield also admits, "In the world of talk shows, there is nothing new under the sun. What makes it different is the host. It has to do with Cavett. Cavett is a little different, a little fresher, without being too heavy."

Yes, with Cavett things rarely get too heavy. He arrives a half-hour late for our scheduled appointment, effectively cutting it in half, but he is so sincerely apologetic and disarming I can't stay annoyed. "The dog walker didn't bring the dog back," he says.

Cavett gives the impression of a man without driving ambition. He continues to act on stage—during the summer he played several weeks of summer stock in *Otherwise Engaged,* in which he had earlier appeared on Broadway—and does commercials and occasional guest spots on television. He speaks of sometime soon performing in a Robert Altman film, as a talk show host. But he seems to have no urge to write, direct, or star, like his friend Woody Allen. "He doesn't have the taste to be a developer and entrepreneur," says Porterfield.

Cavett asked his old professor Paul Weiss, "Here's an easy one for you. What is the root of all evil?" And Weiss answered, "It is misconstruing who you are, ignorance of yourself." Weiss noted that both Socrates and Freud had taken as their motto, Know Thyself. It happened that with several of his guests during the week I visited the show—with Weiss, with Singer, with Sendak—the subject of psychoanalysis came up. I asked Cavett why that pattern had occurred, and if he had experienced psychoanalysis.

"In two of three cases it was in the notes," Cavett explains, referring to the notes on guests the segment producers prepare for him. As for being in psychoanalysis, his answer is, "No, but like Weiss, I sort of wish I were." Characteristically, he turns the subject into anecdote. "One of my favorite moments was when R. D. Laing was once on the show. He said, 'You're going to be saying I'm paranoid in a moment.' And I replied, 'No, I'm not. You're just imagining it.' "

No more than Porterfield can Cavett articulate what makes him a successful talk show host. "They're always raving about me in Europe," he says, by way of roundabout response. "There's something they can't master. I don't know what it is." German television "even had me over to reveal the secrets of the talk show." Cavett seems to be a little less fully confident in his abilities than people like Porterfield who organize his work and constantly push him. Remember Cavett, the perennial student.

"Weiss made an interesting remark," Cavett says, quoting the philosopher: " 'If Dick asked me things that interested him, it would be more interesting to people.' " He mentions another friend who is "always encouraging me to trust my instincts." One radical piece of advice was not to have any staff. "It's a little impractical," he muses.

So Dick Cavett continues his liberal education on PBS stations. It may never have occurred to him to invite such guests as Maurice Sendak or Isaac Bashevis Singer, and he may have chafed at doing the homework the segment producers assigned, but when the red light on the camera goes on, he is bright, curious, prepared. And because of Dick Cavett, we continue our own liberal education: People talking about work and art and ideas can still matter more on television than how many people tune in to watch them.

12. Growing Up with Joan Ganz Cooney

"Props, we need some more cigars," calls the stage manager on the Children's Television Workshop set. "And some matches, please."

A prop-man hurries over with a fistful of stogies. "Who needs a cigar?" the stage manager cries. Hands go up among the extras, and the prop-man tosses cheroots left and right. "Those of you with cigars," the stage manager says, "let's keep them up high."

A towheaded child among the extras asks for a cigar.

"When you're over three-feet tall," the stage manager replies.

"I am," the boy says softly, but his words are lost in the bustle of preparing the scene for taping. Close, but no cigar.

A "Saturday Night" parody of a day in the life at "Sesame Street"? No, it's the set of an episode for Children's Television Workshop's first dramatic program for adults, "The Best of Families," for broadcast on public television. The eight-part series traces the lives of three fictional families in New York from 1880 to 1900. One is a large Irish immigrant clan, another respectable middle-class, the third aristocratic.

The scene requiring the cigars occurs in a show near the end of the series. It is New Year's Eve, 1899, in an Irish saloon. The immigrant neighborhood has changed, Jews and Italians are moving in amid the Irish, and the Raffertys and Fitzpatricks have turned the saloon's old family room into a moving picture hall, showing the newcomers flickering images of a boxing match and of "Little Egypt," the belly dancer. Extras are dressed in bowlers, rough caps, and the coarse woolen clothing of the turn-of-the-century urban poor.

"The Best of Families" shows a wide spectrum of late 19th-century American social life. In other strands of the series, Sarah Baldwin, daughter of a Brooklyn minister, seeks both marriage to an architect and a career as a newspaper reporter; and Teddy Wheeler reverses his family's declining fortunes by an ambitious rise to prominence in the worlds of politics and finance.

Children's Television Workshop has its own ambitions at stake in "The Best of Families." The workshop is so concerned that viewers may be confused by its name and reputation as the leading programmer of daytime children's shows, it has asked the press to describe the production company only by the initials CTW. Though the Workshop's "Sesame Street" and "The Electric Company" have accounted for nearly half of public television's audience in the past, the Workshop has yet to prove itself as a prime-time producer—its one previous major effort, "Feeling Good," a health series, got poor notices and was discontinued.

Why does Children's Television Workshop need to prove itself on evening television? It has already earned an honored place in television history by creating two important programs for children and by demonstrating that educational lessons can be successfully presented in an entertaining television format. And that, of course, is just the point. As public television enters its second trial of soul-searching, by means of a new Carnegie Commission study, the Workshop stands out as the greatest triumph for public television in the past decade. For the people at CTW, it seems reasonable, even imperative, that the Workshop expand into prime-time public television.

But that's easier said than done. CTW is not CBS. It started as a small experiment with a well-defined aim, to produce a television program that could effectively teach preschoolers. A nonprofit enterprise, CTW depends for its funds on private foundations and corporations, the national endowments, and the Corporation for Public Broadcasting. Most grant-based projects complete their task and fold up. CTW, instead, has grown beyond the bounds of an experiment, beyond the bounds of children's programming—but not yet beyond the bounds of a grant-supported institution. Its growth has increased its need for outside funding, and it has still to decide which of its many possible roles in public television it can best perform.

"I think that now we feel that we are primarily a production company with a mission," says Joan Ganz Cooney, the founder and president of the Workshop. "The Best of Families" was a concept that grew out of the sense of mission. Naomi Foner, a young producer on "Sesame Street," had an idea for a dramatic series on American history. She asked for an opportunity to spend a year in London with the BCC, studying its successful productions of historical dramas. Somewhat to her surprise, it was granted.

"This was an old idea of mine," says Foner. "It grew initially out of watching 'The Forsyte Saga.' One could use entertaining and noncondescending material to present American history on television. Whatever American history I've learned is from media exposure, and it's probably all wrong. You start with the emotions and your head follows. What we wanted to do had never been done without distortions. We weren't going to sacrifice either accuracy or drama."

During her year at the BBC, Foner discovered that the British made no claim their historical dramas were accurate history, just the opposite of what she aimed to do. She also learned that British public television benefits from a regional system that trains talent, particularly writers, over many years, and from owning and constantly using its own production facilities.

The Workshop, with neither of these advantages, decided to go ahead with Foner's proposal anyway, but it did make one concession. Concerned about its staff's lack of experience in dramatic production, the Workshop brought in Ethel Winant, a long-time CBS producer, as executive producer on "The Best of Families," and hired among its writers and directors a number of commercial television hands.

Children's Television Workshop got its start when another young woman got a chance to take some time off from her duties and to study the possibilities for innovative television. The year was 1966; Head Start and other federal programs were trying to improve preschool learning opportunities for disadvantaged youngsters. Joan Ganz Cooney, from Phoenix, Arizona, a producer of public affairs documentaries for WNET, the New York public television station, received a $15,000 grant from the Carnegie Corporation of New York to spend three months talking with educators and television people about television for preschoolers. Those talks led to the founding of CTW, to the creation of "Sesame Street," and finally to the rise of Joan Ganz Cooney to one of the most important positions in American public television.

She did not, of course, do it alone. Her primary aegis was Lloyd Morrisett, chairman of CTW's board of trustees. As an officer of the Carnegie Corporation, Morrisett initially funded Cooney's research. Moving to the presidency of the John and Mary R. Markle Foundation, he provided additional support. Half of the Workshop's original funding came from the Department of Health, Education and Welfare, the other half from the Ford Foundation, the Corporation for Public Broadcasting, and the Markle Foundation. Morrisett still serves as Cooney's adviser and sounding board. They try to lunch together once a week.

Cooney presides over the Workshop from a large corner office overlooking Lincoln Center. Around her are Muppet dolls and street signs in foreign languages from the "Sesame Street'" international coproductions, symbols of CTW's extraordinary worldwide impact on young people's learning and play.

"It was clear it would work," Cooney says. "When people ask, 'Were you surprised?'—I was never surprised. I knew it would work. It was money working for money. We knew we were going to have outreach activities.

"I did not understand that success has its own problems. You're in the production business. After six months Lloyd Morrisett says, 'Big Bird can't fly on one wing. Plan another show.' Success was not simple. Once there was the idea of doing a show and self-destructing. But instantly we were in the growth business."

"There was," Cooney adds, "an incredible amount of international attention. Mike Dann wanted to leave CBS, and he started the international division. The phone started ringing off the wall. It was a revelation of some kind."

"Sesame Street" was also a revelation of sorts to its viewers. And that, too, was not exactly serendipity. From the start, a collaboration between educators and television producers shaped the program. They researched program design, and they evaluated the results among children. They wanted to hold a child's attention to learning material the same way a clever commercial can rivet the eye. Dr. Edward Palmer, the Workshop's vice-president for research, and his staff developed a device to show children images which might distract them while they watched "Sesame Street" on a television screen. This way Palmer discovered which program segments held children's attention and which lost out to the distraction, and revamped the program accordingly.

Later, children were surveyed to see if they grasped what the programs tried to present. " 'Sesame Street' was held to the test: Did they learn?" Joan Ganz Cooney says. "In the beginning we tried to teach the alphabet and numbers in a single season. We now know we have the kids for some years, from one-and-a-half on."

Now in its ninth season on PBS, "Sesame Street" not only has American children for some years, it has children from more than fifty countries around the world. CTW sends abroad both the English-language version and an adapted series designed for foreign-language sound tracks. It is co-producing "Plaza Sesamo" in Latin America, "Vila Sesame" in Brazil, "Sesamestrasse" in West Germany, "Sesamstraat" in Holland and Flemish-speaking Belgium, and the latest, "1 Rue Sesame," in France.

While "The Best of Families" was shooting at the old Twentieth Century-Fox studios in West Manhattan, a few blocks away television personnel from Kuwait were visiting Lincoln Plaza, preparing an Arabic-language pilot of "Sesame Street."

"Sesame Street" has grown and changed with American society over the past decade. It has given more attention to the role of women, to cultural diversity in the United States, to health and nutrition practices. But the program still rests on a solid foundation of imaginative and slightly wacky humor. "The secret," Cooney says, "is to be able to write comedy for four-year-olds."

I tuned in "Sesame Street" the other morning and encountered a tanklike machine with bulging eyes that goes by the name of Sam.

Sam was professing his love to Gordon's Volkswagen, Ramona. "Why can't a machine love another machine?" asks Sam, and proceeds to sing a love song to the car. Sam departs; Gordon returns and finds Ramona's motor running and signals flashing. He scratches his head, knowing full well he wasn't the one who turned on his car.

Then there was a filmed scene about a Hispanic boy who spills the milk his mother brings to the breakfast table. How did the milk get spilled?, his mother asks, and the boy lets his imagination hold sway—a robber did it, he says, and the film pictures the scene of a masked man spilling the milk, stealing a cookie, and departing. No, he says, it was a wrecking crew. No, it was a gorilla . . . a baseball player . . a herd of elephants. Each story is visually documented. "Actually," he finally admits, "I spilled the milk."

There's a message in that little tale: best to tell the truth. But oh such pleasant fibs! I suspect I have the sense of humor of a four-year-old.

In 1974, when the A. C. Nielsen Company did a rating survey on public television, "Sesame Street" drew a larger share of the public television audience than all of its evening programs added together. The next most popular public television show was CTW's other program for children, "The Electric Company." But it now appears that the number of children watching the two shows is declining.

"Audiences for the children's block of afternoon public television have been falling off because of repeats," Cooney says. " 'Sesame Street' should be revised enough so you could publicize what you were doing. The mothers of preschoolers have not been bombarded with publicity on education. Independent stations are plotting against 'Sesame Street' with game shows. The independents are programming against the block and the block has not been refreshed. If a show feels old-fashioned to a three-year-old, then a three-and-a-half-year-old is not going to watch it."

The answer to the problems of "Sesame Street" probably can be given in one word: money. The Workshop is hardly unique in the world of public television in its constant quest for funds, but few other nonprofit organizations have moved so fast to grow and diversify as a solution.

The Workshop has spun off two subsidiaries which operate in commercial broadcasting fields. Palm Productions, Inc., produced a dramatic special for NBC, "Beauty and the Beast," with George C. Scott and Trish Van Devere, and has been developing a half-hour situation comedy for prime-time network consideration. CTW Communications owns a radio station in Ventura, California, and has holdings in the cable television franchise for Honolulu, Hawaii. The Workshop hopes that after-tax profits from these enterprises can be channeled back into nonprofit programming, thus lessening dependence on outside funding sources.

New projects, meanwhile, generate new funds. At NBC's request, the Workshop is producing a series of three pilot "health notes," patterned after the Bicentennial minutes, for broadcasting as prime-time spots. It is also developing a third major program for children, a series intended to serve as an introduction to science for eight- to twelve-year-olds.* ("The Electric Company" has completed its projected five-year production run and is on repeat broadcast cycle scheduled to run through 1980.) And the Workshop has also acquired rights to Richard Kluger's *Simple Justice,* a history of school desegregation, which it plans to dramatize in a series for PBS.

This volume of work requires tireless efforts to pay for it. CTW was not successful, for example, in getting outside sources to cover the entire budget of "The Best of Families." The National Endowment for the Humanities, the Mobil Oil Corporation, and the Corporation for Public Broadcasting put up most of it. CTW covered the rest. "There's always been a tendency in public television to be stingy," says Cooney.

Cooney wants the federal government to shoulder more of the burden of paying for public television. "Britain, Russia—both free societies and Communist societies—support the arts. Japan gives part of its gross national product to public television. Here we have the legacy of the frontier mentality—Puritanism and freedom together. We have vast government interference in every part of our lives, and we keep trying to keep them out of the arts."

She even proposes a third federal endowment to go alongside

* The science series, entitled "3-2-1 Contact," began its run on public television stations early in 1980.

the arts and humanities endowments—a National Endowment for Children's Television. Cooney is definitely not among those who think that Children's Television Workshop has provided a sufficient answer to the need for quality children's television programming. Her major disappointment is that CTW has, in her view, made little or no impact on commercial television.

"We have been pronounced a historic force," Cooney says, "but that's not quite true. It didn't make a hell of a lot of difference in broadcasting. I had wanted us to have competition that would lead us to take 'Sesame Street' off the air. I thought we'd have competition. There's 'Captain Kangaroo.' No one else has anything.

"After school—it's tokenism," she adds. "It's incredible. 'Sesame Street' and 'Captain Kangaroo' are the only two daily shows for the kid audience. Television viewing by children goes way up in the summer. You could put on a kid's show at 10 p.m. and have a great kids' audience."

Television has yet to scratch the surface of what it could be doing for children, Cooney believes. "We don't put on good fairy tales for kids. We don't educate kids in our own myths. It's a great opportunity for the country."

Joan Ganz Cooney looks at the state of children's television and sees a pinchpenny public television, indifferent networks, ungrasped opportunities. Others look at CTW and see the successful creation of a ground-breaking organization by an exceptional woman leader. Joan Ganz Cooney is in demand. Foundations seek her counsel. Universities give her honorary degrees. Corporations ask her to serve as a director. Organizations present her with awards and appoint her as a trustee. She sits on the board of, among other organizations, the American Film Institute.

A few days before I talked with Joan Ganz Cooney, she had been photographed for a fund-raising effort by a women's organization. In her early years at the Workshop she had kept at arm's length from the women's movement (although she had been a charter member of the National Organization for Women). Now she feels more a part of it.

"I felt initially the women's movement had a kind of anti-male flavor that was going to hurt the work I was trying to achieve.

Lloyd Morrisett, and Mac Bundy of the Ford Foundation, were willing to sign up with an inexperienced woman leader. I have stayed very, very far from controversial issues. If I want to get into politics I should get out of the Workshop. Our work is a political-cultural-social statement."

It bothers Cooney when one of those statements gets rejected, like "Feeling Good," the health series that was canceled. "We kept thinking it was a failure. We believed what we read in the newspapers." She adds, "Lloyd Morrisett now thinks we misunderstood—our expectations may have been totally unrealistic. That show may have been quite at the top."

People will act on information about health habits and behavior, Cooney thinks. "People quit smoking, as I did for thirteen years," she says ruefully, holding a lighted cigarette in her hand. Whether people absorb the information or not depends on how it is presented. You have to find a way to make things vivid, unforgettable.

She recalls a commercial that goes back, oh, a good twenty years, and may only be familiar to viewers from the Middle Atlantic states. It was for Piels beer, brewed in Allentown, Pennsylvania, and the commercial featured animated cartoon characters named Harry and Bert, the Piel brothers. It was a big hit at the college I was attending at the time. It was a big hit with Joan Ganz Cooney, too. "When it started I knew it was striking," she says. "I dreamed about it every night."

Could Harry and Bert Piel, those beer purveyors, be the guardian angels of Children's Television Workshop? Come to think of it, there are a lot of little men on "Sesame Street" pushing the letters *J* or *L* with an insouciance reminiscent of good old Bert and Harry—and isn't there a "Sesame Street' duo called Bert and Ernie? After all the research and government programs and foundation grants that went into the making of CTW, it's pleasant to think that "Sesame Street" and the Workshop may ultimately be products of Joan Ganz Cooney's dreams.

13. Good Morning, Lanesville

To reach America's smallest television station you go through Phoenicia, New York, and drive five or six miles up the valley road, past houses with cows grazing on the lawn, till you reach the big white farmhouse on the left. There, in the heart of the Catskill Mountains, live the people of Media Bus, proprietors of Lanesville TV, Channel 3. For the past six years they have been broadcasting more or less regularly on Saturday nights at seven to Lanesville, New York, and environs—an effective radius of three miles. Lanesville TV numbers its potential audience around three hundred. Nothing to make William S. Paley quake in his boots just yet. But wait: Lanesville TV may merely be a little ahead of its time, a straw in the wind, a harbinger of television's future.

You'd be hard put to find Lanesville, New York, on the map of American television. But on the map of American *video,* it stands out as one of the capital centers, alongside Manhattan, Southern California, the San Francisco Bay Area, Seattle, Boston, and a few other places. Back in the medium's infancy, video and television used to be interchangeable words, as in Captain Video, the space pilot of an early children's adventure show. Now, however, their meanings have been split asunder, and video is used in a narrower sense to denote the making of art works and documentaries on half-inch tapes.

For most of America's television watchers, the video map is terra incognita, but it's getting more visible all the time. You can find video in a handful of museums, a gallery here and there, occasionally on PBS. Video may be the most dynamic and diverse form of art and communications in America right now, though it has hardly begun to find ways to reach a larger public.

Videomakers are the people with the portapaks. They began to appear in the late 1960s with the arrival from Japan of lightweight portable cameras and recorders using half-inch tape. Some filmmakers were attracted to video. Others came to it from the visual arts, painting and sculpture, and yet other videomakers were *sui generis*—born as communicators with the new medium. Of videomakers there are many kinds: documentarians, conceptual artists, mixed-media performers, creators of video environments. And many of them find their way to Lanesville, where they can broadcast their work to an audience for the first and perhaps the only time.

Lanesville TV is hardly the complete word on video. But it has probably the clearest vision of relating video to you and me, the people who watch television. So when I want to learn something about video it is to Lanesville that I make my way, driving through Phoenicia and up the valley road, past the grazing cows, to the big white farmhouse, Maple Tree Farm. The day I pick is a lucky one: After a hiatus of nearly four months, the longest gap in programming since the broadcasts began, Lanesville TV is going back on the air.

The main reason is that its viewers want to see the tape of "Harold's Bar Mitzvah." Harold is Sam and Miriam Ginsberg's grandson. When Sam and Miriam and a few other Lanesville folks went down to Long Island for the big event, Bart Friedman of Media Bus and Lanesville TV went along with his portapak. After producing more than two hundred fifty programs and making over five hundred tapes, the people of Media Bus know pretty well what their audience likes.

Bart Friedman is an intense, bearded man in his early thirties. He has an aggressive vein of humor that quickens his energy as it leavens his seriousness. "People like to see themselves and their families," he says, but they're interested in other things, too. When the Chinese pandas arrived at the Washington zoo, Lanesville TV went down and covered it, slanting the tape, of course, to the Lanesville audience.

The usual broadcasting format is a combination of live action and tape. On the live segments the farm's telephone number is frequently displayed, and people are invited to call in. Sometimes callers just say, "The reception is good tonight," or, "You're coming in clear." Other times they talk about local events, gossip, things lost and found. Once when Lanesville TV premiered dance tapes that were later to be part of a PBS Twyla Tharp special, a neighbor called in and complained, "Take off that dancing, get on with the show."

The closest analogy to Lanesville TV may be your local weekly newspaper. But even that doesn't fully tell the story of what the station is trying to do. It's like a local newspaper that each week publishes an original short story by Saul Bellow or an original essay by E. B. White. It's like a local newspaper when the readers write a few of the stories and comment editorially about others. Lanesville TV may be like no other form of communication before it.

Dare To Be Boring may be the most apt motto for the programs, Bart says. The station has put on political shows, cooking shows, sports programs, art, and abstract video. "You have to watch us for a long time to appreciate us," he says. "We deal with our own conception of what people look like."

When the videomakers first arrived in the valley they were regarded as long-hair hippie intruders, but, after six years as neighbors, an intimacy has developed that reflects itself on tape as well as in life. Bart Friedman didn't just record Harold's bar mitzvah, he created his own conception of it on tape—and incorporated in it the conceptions of his subjects as well.

"There's very little sharing of reality," Bart says. "Video and this kind of broadcasting allow us to share kinds of inner space—docufantasy, call it—that only video can do." By sharing "inner space," Bart means that the Lanesville videomakers give up complete control of the image, allowing the subject's dreams, fantasies, imaginings to play a role in creating the work of video art. "Docufantasy" means almost literally documenting the subject's fantasy.

The most famous videotape to come out of Lanesville, "Harriet," made by Nancy Cain in 1973, is the quintessential "docufantasy." It is a story about Harriet Benjamin, a Lanesville neighbor. Harriet is shown in the everyday setting of her home, part trailer, part shed, doing the household chores—making beds, washing dishes, hanging clothes, caring for her four kids and husband, with country music whining from a radio in the background. But every so often there are flash-forwards interspersed among Harriet's mundane tasks, scenes of Harriet escaping, laughing, driving away.

Finally Harriet's fantasy takes over the tape. She jumps in the car, slams the door, and roars away, crying out, "No washing, no ironing, no cooking, nothing. I'm sick of Lanesville. I want to see something different. Good-bye Lanesville. I've had seventeen years of it and that's enough." She sings "Roll Out the Barrel, We'll Have a Barrel of Fun," as the car races down the valley road.

"When she got into the car, I didn't know what she was going to do," Nancy Cain says. "It was her own idea, her own self." Perhaps Harriet didn't want Nancy's tape to be merely a record of her own rather dismal-seeming daily reality. She used the videomaker's willingness to follow her with the tape rolling as an opportunity to present a different self. Nancy made room for Harriet's "inner space" on the tape, and Harriet seized the opportunity.

A more communal work in the same vein is "Frank 'The Fist' Farkle vs. Rocky Van," a tape of a boxing match staged to raise money so the Hunter-Tannersville Fire Department-Rescue Squad could buy a new truck. The boxers, their entourages, and the spectators are all willing creators of their own "mythic moment." Like primitives imitating the gestures of the gods in their rituals, the participants become role players, outsized selves, copying the rituals of the prizefight and the eccentricities of boxers like Muhammad Ali—movements and phrases observed, of course, on television.

About forty-five minutes to air time, the farmhouse begins to bustle with activity. Counting Bart Friedman and Nancy Cain, there are ten people working on the broadcast, some residents, some visitors. Bill and Esti Marpet, New York videomakers, have been staying at the farm editing a video documentary on the running of the bulls at Pamplona, which they made with the support of the TV Lab at WNET, the New York public television station. Part of "Running with the Bulls" will be shown on the broadcast. New York filmmakers Maxi Cohen and Joel Gold are there, too. Maxi and Esti will operate the cameras, Joel will play the piano, Bill will assist Parry Teasdale, Media Bus's technical writer, at the controls.

A Victorian parlor is transformed into a broadcast studio. Cables are laid from the mikes and video cameras in the parlor to the control room. Sound is checked, cameras focused, control room set up, as nearly everyone gets pleasantly stoned. More than one person, observing the maze of cables crisscrossing the floors of several rooms, remarks, "Spaghetti City," a reference to a book on alternative video written by Media Bus in a prior incarnation under the name Videofreex.

At 7:01 p.m. Parry calls out, "We're on the air," and the show begins with Bart playing saxophone and Joel at the keyboard. They improvise some lyrics—"If you'd like to watch over here, just come through the door," Bart sings—and by 7:04 "Harold's Bar Mitzvah" is rolling. The tape is twenty-two minutes long, giving several people time to walk over to Sam Ginsberg's to check reception. "It's clear at Sam's," they report. "You can see and hear."

People sit and watch the tape on monitors. On the tape, Sam and Miriam are talking to Bart, unseen behind the camera, about the clothes they will wear to the bar mitzvah. They get in a car, and you see a shot of the valley road.

"Is that from 'Harriet'?" Parry asks in the control room.

"He dipped into the video bank and took out a withdrawal," Bill Marpet says, "some fireworks and a traveling shot." The fireworks, which opened the bar mitzvah tape, are from Bill's Pamplona footage.

Bart appears on the tape, mike in hand, like John Chancellor at a political convention. "Here we are at the bar mitzvah," he says. "Thirteen wonderful years have passed."

Soon Bart tells the control room, "This is the last cut; it will last about a minute." Everyone jumps back up to resume live broadcasting. A smooth cross-fade carries the viewers from the bar mitzvah to Joel at the piano. "Our phone is broken tonight so we can't take calls," Bart tells the Lanesville audience. He explains that it's their first broadcast in some time, that they plan to break with the old regular Saturday evening format, programming at new times—Midnight Lanesville, Good Morning, Lanesville. They'll call viewers and put up posters in the stores to let them know.

"Remember," Bart says, "we're not listener-sponsored, not commercially sponsored. This is *my* TV."

January Time, a Maple Tree Farm resident, sings. Over her song Bart says to the viewers, "How could you turn it off? I would watch it for thirty hours." January finishes, and Bart introduces the Marpets' Pamplona tape. "It's a true test of masculinity, not like anything we've had in Lanesville," he says. "What kind of test of masculinity is there in Lanesville? None that I know of, unless it's drinking and swearing."

The Pamplona tape rolls for twenty minutes or so, and they decide to take a break. Bart comes back on live and says, "We wanted to give you folks a chance to go to the bathroom, get a beer," and after a minute more the tape resumes. A little after half-past eight they decide it has been enough. Bill and Parry work out a program close with shots of exploding fireworks and the sound of a cheering crowd from the Pamplona tape.

"My goodness, we haven't lost our touch," Bart exults. They shut the power, pack away the cables, cameras, mikes, and all go in to dinner.

What has this little adventure in homemade broadcasting to do with the monolith of commercial television, the few dozen Lanesville viewers with the tens of millions that CBS, NBC, and ABC are fighting over, like *Star Wars* spaceships in a cosmic rating battle? More than you might guess. For it was one of those networks that helped get alternative television off the ground, that hatched its small but tenacious rival, as they used to say back in the 1960s (borrowing from George Orwell), right in the belly of the whale.

The year was 1969, the time of *Easy Rider* and the Woodstock festival, and not a few people at the networks were wondering what would become of them if the younger generation passed the crucial eighteen to thirty-nine consumer years stoned on grass and Jimi Hendrix. At least one executive, Michael Dann of CBS, took the precaution of putting an assistant on the case. Could he come up with a concept for a show that would appeal to those mysterious but free-spending youth? One of the assistant's assistants was Nancy Cain.

Nancy remembers those days as the "myth times." Does it feel like a myth, from the angle of Lanesville, that she was there, walking the corridors of Black Rock? Nancy and her boss, at any rate, heard about some videotapes people had made at the Woodstock festival. They took a subway to a Lower East Side loft and met the videomakers, David Cort, Parry Teasdale, and Curtis Ratcliff, and they saw the tapes.

" 'My God, I'm saved,' " Nancy remembers saying to herself. "I was really happy to see these types and their crazy, drugged-out, fabulous tapes. We were saved in the nick of time." They had found the people who could produce a pilot that reflected the style of the counterculture.

"We wanted to make some kind of TV that showed what was really happening, showed real life, that didn't really end, had no format—some kind of channel you could tune in and out of at will." They rented a loft in the SoHo district of New York. The group became a magnet, drawing people whose lives were changing in the political and cultural ferment of the time.

Soon the pilot was ready. They set up a studio in one loft, and fed the picture to Mike Dann in another. A musician, Buzzy Linhart, performed live, interspersed with taped segments on counterculture subjects like alternative education. "It had a spirit of chaos, in a certain sense," Nancy says. She recalls that Mike Dann said something unprintable and added, "This is five years ahead of its time." The CBS experiment was over.

But the group lived on, and Nancy stayed with them. They became Videofreex, took part in the revolution of portapak documentary video that spawned such groups as TVTV (Top Value Television), now of Los Angeles, and Ant Farm of San Francisco. With funds from Abbie Hoffman they bought a transmitter, a modulator, an amplifier. They were ready to begin broadcasting.

One day in 1971 they were editing videotapes of the May Day demonstrations in Washington, when Con Edison turned off the electricity. That same day they signed the lease on Maple Tree Farm. "People wanted a garden," Nancy says. "It clears your mind."

Upstate, one hundred twenty-five miles from New York City, Videofreex became Media Bus—partly because they supported themselves by going around demonstrating video equipment for New York State. Grants from the New York State Council on the Arts also enabled them to establish a media center where videomakers could do post-production work. "We have three windows in our control room, overlooking beautiful mountains," says Bart. "It's quiet, you can work all day, there's a resident engineer, we charge no money." Among the artists who have availed themselves of this remarkable deal is Lily Tomlin, who re-edited video segments of her 1977 stage show there.

No other group of videomakers has yet to follow the path of Media Bus into independent local broadcasting. TVTV got its start in 1972 making iconoclastic, behind-the-scenes, video-vérité documentaries about the national political conventions, and syndicated the programs nationally on cable and local broadcast stations. Since then many of its documentaries—like "The Lord of the Universe" (1974), about a meeting in the Houston Astrodome of followers of the Guru Maharaj Ji, and "Gerald Ford's America" (1975)—have been funded by public television and broadcast by PBS.

Ant Farm, whose roots in other forms go back to 1968, makes what may be called anarchist-dadaist videotapes, often based on its own staged happenings, that satirize American fetishes like the automobile and television itself. Its "Media Burn" (1975) is both an event and a tape of the event and its coverage by local news programs. On July 4, 1975, Ant Farm set fire to a stack of some three dozen television sets and drove a car through the video pyre. Before the burn, a John F. Kennedy look-alike arrived to deliver a Kennedyesque Independence Day oration on the dangers of the American addiction to television. "Haven't you ever wanted to pur your foot through your television screen?" he asks, and concludes, "The image created here shall never be forgotten."

Neither TVTV with its national programming successes, nor Ant Farm with its vivid comic deflations, have done what Lanesville TV has—created a working, if modest, model of a system of television delivery different from the one we have now. "We're learning," Bart Friedman says. "This is real. This is war. Someday we're going to take over CBS." And he is only half-joking.

Shigeko Kubota has no visions of broadcasting power. She is a maker of video environments and is video curator at the Anthology Film Archives in New York, where she programs weekly presentations of artists' video works. To reach her loft-studio you take the BMT to Prince Street and wend your way past workmen loading trucks to find a gray, cast-iron manufacturing building on Mercer Street. There, in a fifth-floor loft, bright with the light from interior air shafts, Shigeko Kubota lives with her husband, Nam June Paik, perhaps the most renowned of contemporary videomakers, and creates her video sculptures.

Video sculpture, like Lanesville TV, is another way of reshaping our experience of what television is and can be. It makes us realize television can have a spatial dimension as well as being an image. It makes us grasp concretely what we already know abstractly, that the video images come from a signal, not a film, and can be repeated infinitely.

All this is easily said, once you've seen the work, but how does the artist get the idea, before anyone else has thought of it? In the case of Shigeko Kubota, as in Lanesville TV's, necessity was at least in part the mother of invention. She had made a tape, "Video Girls and Video Songs for Navajo Sky" (1973), contrasting city images with the life of the Navajo, but she could not interest any broadcaster in putting it on the air.

Shigeko is an ebullient woman, about forty, who speaks a rapid-fire English with some of the syntax of her native Japanese.

"They didn't like it," she says. "It was too intimate for public television. I was rejected by broadcast video. Where could I go? I would make my own video environment."

Finding no place in the existing structure of television, she decided quite literally to make her own. And it was to be a challenge to conventional ways of experiencing the medium. "I wanted to make a realization of video," she says, "to go beyond the strip of the videotape. Editing, chop, chop, chop, I don't like. Videotape was a subject; I would make it an object. I just make a more natural video environment. Video time is real time. I make a marriage of subject and object."

When the Kitchen Center for Video and Music in SoHo, one of the few organizations actively fostering video art, invited her to mount a show, Shigeko chose to build it around her longtime interest in the legacy of Marcel Duchamp. She had published a book, *Marcel Duchamp and John Cage,* made up of photographs she had taken of the chess match the two artists had played with the board wired for sound. Now she took videotapes she had made of Duchamp's grave (from her "Europe on ½ Inch a Day" of 1972) and built a structure to present them.

Called "Gothic Video Tower of Marcel Duchamp's Grave," it took the form of an austere vertical wooden structure enclosing tightly spaced video monitors. Along the floor in a direct line she places narrow mirrors. The sculpture reaches from floor to ceiling in whatever space it is housed. Currently, standing in her own loft, it has eleven monitors—all showing the same tape of Duchamp's grave.

Continuing her exploration of Duchampian themes, Shigeko built a four-tiered wooden video sculpture called "Nude Descending a Staircase." Each step of the piece encloses a monitor and each monitor plays a simultaneous image of a nude woman descending a staircase. Thereafter came more video sculpture in the Duchamp vein, including one called "Window," a wooden construction of a window frame with eight panes, behind each of which sits a monitor playing images of snow. Besides at the Kitchen Center, her video sculptures have been exhibited at the Long Beach Museum of Art, And/Or in Seattle, the Everson Museum in Syracuse, New York, at galleries and museums in New York City, and in Germany.

Shigeko Kubota was a sculptor in Japan when she was invited in the mid-1960s to join a new dissident art group in SoHo called Fluxus. It contributed street art, happenings, and destructive art to the New York scene. "I stopped making art and made street art," she says. That lasted until the end of the decade. Then she felt the need "to make a structure in my art after anger art and street art." The search for structure led to the video portapak, to video sculpture, and finally to seeking a showcase for all types of video artists.

"In 1974 I knocked on the door of Jonas Mekas," she recalls, referring to a central figure in the world of independent cinema, the head of Anthology Film Archives. "I said to Jonas, 'Why don't you have video at your theater?' He had a kind of delicate attitude toward video." Nevertheless, he agreed, and the twice-weekly showings were established. Video now sometimes draws a larger audience than film at Anthology archives.

"Video is not yet established, not yet competitive," Shigeko says. "I think in film the possibilities are more limited. You cannot compete with Stan Brakhage or Hollis Frampton. I feel sorry for young filmmakers. Hollis Frampton is there like a stone. You can't move."

So the aspiring videomakers submit their wares, and Shigeko Kubota watches hours and hours of it. "I get bored," she says, and one thanks her for her candor. Still, most of it gets programmed. "Unknown people need an environment," she says. "Video is not settled yet. I don't want to select just yet."

Video art is not yet a decade old. Although it is far from being established, some memorable work has already been accomplished. The highlight of my own experience of video art came at a show by Nam June Paik several years ago at Bonino Gallery in SoHo, "Fish Fly in the Sky." I entered a darkened gallery and was greeted by a sign, Come In and Lie Down. As my eyes adjusted I saw gym mats on the floor, and found one to stretch out on. Looking up at the ceiling I saw perhaps thirty monitors installed there, playing different cycles of the same tape loop, tuned to different color tones—full color, blue, red, yellow, green, even black and white. The tape showed goldfish swimming in a tank, and shots of blue sky, clouds, skywriting. An audio tape played soft surf sounds. It was art, it was meditation, it was religion, it was, for some, a chance for a wonderful nap.

"Video will be the main art of the 1980s," says Shigeko Kubota. You will find it more and more in galleries and museums and, if Lanesville TV has any influence, on broadcast television as well. Whether it may change the face of American television remains to be seen, but we can hardly go wrong with a television experience that offers us the neighborliness of Lanesville TV or the invitation, as in "Fish Fly in the Sky," to come in and lie down.

Epilogue: This evening of broadcasting may actually have been the final one for Lanesville TV. Not long after, the people of Media Bus gave up Maple Tree Farm and moved to Woodstock, New York, where they continue to work on video projects.

PART THREE

Television Criticism

14. On Television Critics and Criticism

This is an essay about a subject that does not exist: television criticism. There are, of course, people called television critics. I happen to be one of them. But it is one thing to pursue a line of work—in this case, writing about television programs—and quite another to build a body of work that helps to define the form, meaning, and direction of the field it surveys. Television critics are the hod carriers of criticism, who have yet to figure out how to lay their bricks and mortar side-by-side. When was the last time you read a review that said anything enlightening?

What I have in mind is a type of television criticism that makes a difference to the television audience. A television criticism that helps its readers understand why a program that works, works, and why one that doesn't, doesn't. A television criticism that notices what creators do, not what networks hype. A television criticism that takes television seriously as a cultural form, rather than as a monster threatening to destroy civilization, or, perhaps worse still, as a playing field for programming superstars.

It can happen. In Britain, television critics review programs after they've aired, much like a theater, dance or music critic, of necessity, writes a review after a performance. Some readers might regret having missed a show the critics praise, but those who did watch it can compare their judgment with the critics'. And if the criticism is of sufficient interest, any reader, viewer or not, can gain some general perspective on the medium.

You may think that I am making that most common of critic's mistakes: elevating the importance of the critic above all else. It certainly has happened elsewhere. A half-century ago H. L. Mencken wrote a satiric little essay about a new field he called "Criticism of Criticism of Criticism"—and that was long before the swarms of academic exegetes infested the literary vineyards. And there are movie critics who have become celebrities in their own right, whose readers boast that once they've read the reviews they no longer need to see the movie.

I aspire to neither of those conditions for television critics or criticism. Nor am I quite so utopian as to expect that one bright dawn the creative community will pay attention—*really* pay attention—to what television critics say. Somewhere, sometime, a creator (not in television) said to a critic, "You showed me something about my work I hadn't seen before." But not here and not now. Why? Because if a critic ever did express a constructive idea about a program or about programming in general, where is the time, cash, patience, talent, network approval and ratings cushion to give it a try?

So for now television criticism exists more or less as a preview service. With rare exceptions, television reviews appear before a program is broadcast. Presumably they serve as guides to viewers, calling attention, at least, to worthy programs readers might otherwise ignore.

Our critics, those preview servants, are therefore a disparate bunch, struggling against the tide of new programs that threaten to engulf them, flooded with an acute sense of powerlessness, yet buoyed occasionally by the reminder that they are read, and that readers do care—but generally only when their favorite program has been panned.

Is it any wonder, then, that many of the people who write about the medium seem to dislike it? I want to be careful to make my meaning clear: I don't mean simply that they write negative reviews but that they actually seem to fear the medium.

When they dissect the dramaturgical flaws of a particular work, often quite justifiably, they do it more in anger than in sorrow.

They enjoy taking cheap shots at network nitwits and nabobs. And their anger seems to be more general than specific. It isn't that particular programs are judged by the highest aesthetic standards and found wanting. It's that television is made to stand implicitly for everything—about America, or mass culture, or consumer society, or capitalism—that the critics don't like. Their hostility is against a perceived concentration of power or wealth, a perceived threat against standards of art or intelligence, that they profess to uphold.

I'm not saying you have to love television to write about it. What it takes is a little honesty about oneself and one's motives, and a little perspective on the medium. Some understanding of the history of culture would help as well. People who imagine that everyone was brighter and more literate and better informed and more cultured before television came along to drag us down ought to have access to time-travel.

In the absence of a television criticism as I have defined it, what follows is an assessment of individual critics who write about television. Bear in mind that what I have to say is personal, partial, and invariably prejudiced. Should any of the critics answer back, we will have attained that critics' state of bliss, as Mencken ironically defined it: critics criticizing a critic criticizing critics.

Speaking of bliss, let us begin our survey with Michael J. Arlen of *The New Yorker*. Arlen has undoubtedly the most blissful assignment of anyone who writes about television. He may write when he pleases, about what he pleases, to whatever length he pleases.

Arlen, as can be imagined, is a critic in a different sense from nearly everyone else who writes about television. His pieces sometimes appear weeks after a program has aired. He watches television, so it seems, not in screening rooms or on cassettes whisked across town to him by messenger, but on his home screen, with wife, children, and friends around. He is candid, personal, and often funny in a wacky, self-revealing way. In a sense his reviews are about himself—his reaction to the world he sees on television and its influence on him.

Arlen is almost invariably a pleasure to read. He makes it clear that he deplores television much of the time but is honest enough to admit that it is part of his intimate life beyond his merely intellectual or aesthetic analysis of it. Several years ago, in a piece called "What We Do in the Dark," he suggested that the way we watch television has sexual connotations and, more specifically, masturbatory ones. Hardly anyone else admits to thinking about television that way, except those in the offices of advertising agencies.

Gary Deeb, on the other hand, seems to have decided to make a play for notoriety by emulating the drama critic John Simon, who calls attention to himself by the excess of his diatribes. Deeb is not really a critic, however, though he bears that title. He's actually a television reporter. His columns for the *Chicago Tribune* are loaded with rhetorical flourishes about "the brainless mopes who commit malfeasance by sullying the airwaves with garbage." All are calculated to raise the ire and blood pressure of his readers at the networks. When he writes about national programming, however, he basically just rewrites handouts.

Deeb's most useful contribution to television journalism (as distinct from criticism) is his legwork and reporting about the local Chicago stations. He has set a standard for getting behind the press release in exploring the policies and power struggles of local television—which these days almost invariably means local news coverage—that all too few newspapers have emulated.

A more complicated case is that of Frank Rich of *Time* magazine. As a general rule, *Time* doesn't cover television. It writes about television when there's a hole in the back of the book because no new movies are coming out that week.

Rich is a superior movie critic whose acuity and independence cost him his job at the *New York Post*. The corporate editorial style at *Time* keeps him more securely in check, and this is nowhere more true than on those movie off-weeks when his assignment is television. More precisely, when his assignment is to score points off television.

Like the *New York Times* (on which more, later), *Time* seems to have made a corporate editorial decision that television is to be deplored. Rich's television pieces are intended to make that point perfectly clear; they are put-downs of setups. What does Rich write about? "Delta House," "Backstairs at the White House," "The Corn Is Green," "Ike," "Blind Ambition." *Time* seems dedicated to popping television's more obviously overinflated balloons, but it doesn't take much of a pin, or a penman of Frank Rich's caliber, to do that.

The Bobbsey Twins newsmagazines are not so similar, curiously, when it comes to television. It's true that *Newsweek* did not include television among The Arts in its table of contents (recently changed to eliminate the separation of The Arts from everything else), but it does give television a good deal more space, more frequently.

Perhaps in line with the magazine's listing of television among all else in life that is not "art," *Newsweek*'s critic, Harry F. Waters, takes a more sociological than aesthetic interest in the medium. He knows, for one thing, that television is more than the occasional overblown mini-series based on a best-selling book, so he writes about game shows like "Family Feud" and does broad-based pieces on subjects like "How TV Treats Cancer," pegged to the docudrama "First You Cry."

When he does write straight reviews of "significant" shows, Waters tends to be upbeat in tone and to focus almost exclusively on the acting rather than on such elements as scripting, directing, or visual style. It may be that Waters prefers the sociological approach, as evidenced by the fact that other writers such as Betsy Carter and David Gelman sometimes review major programs.

One magazine that has no ax to grind against television is, quite obviously, *TV Guide*. Because of that, or in spite of it, Robert MacKenzie has managed to maintain an independent voice in his weekly review column that has run through the television season, from September through June, for the past few years. However, looking at a collection of MacKenzie's columns all in a row, as I have done recently, is a slightly disconcerting experience.

There, week after week in MacKenzie's concise, one-page time capsules, are evaluations of programs that dropped from sight almost as soon as the ink on his piece was dry. MacKenzie's version of television is not the good high spots, as given us by *Newsweek*, nor the bad high spots, as given us by *Time*, nor the high spots that lead to interesting social speculation, as given us by *The New Yorker.* It's television mainly of low spots—redeeming, unredeeming, and those that lead nowhere.

It is to MacKenzie's credit that he lets the chips fall where they may; he is under no compunction to like what he sees, even though the remainder of *TV Guide* hardly does anything else with entertainment programming than puff it. On the debit side, he tends to look at programs with a kind of tunnel vision, evaluating them generally within the terms they set for themselves, filling his small space largely with plot descriptions, leaving himself only a paragraph at the end for a bit of distancing light humor.

If writing a weekly television critique is tough work, how about having to do it almost daily? This is the situation of the daily newspaper critics who, unlike Deeb, actually review programs. The most significant newspapers for television criticism are in, respectively, the hometown of television producers (the *Los Angeles Times*), the hometown of network executives (the *New York Times*), and the hometown of politicians and government regulators (the *Washington Post*). They present a study in contrasts.

For most people who work in the entertainment industry, the *Los Angeles Times* is their morning newspaper. Knowing this, the *Times* was careful for years not to give anyone in the business breakfast indigestion. But times have changed at the *Times,* and not every critic in the View section now raves about every movie, play, concert and art exhibit that got made, performed, or displayed from Trancas to Laguna.

When it comes to television, however, Cecil Smith is of the old school. Everything is "fascinating" or "a delight" or, well, it must not be television. Fortunately, all *Times* television writing is divided into three parts. Lee Margulies specializes in television-industry analysis and reports on local television stations, and Howard Rosenberg is the critic in the trenches.

Rosenberg actually watches almost every pilot that airs. Perhaps more than any other person now writing about television, he can say, "I saw it, I was [in a manner of speaking] there." Rosenberg certainly does not find everything he sees delightful and fascinating. He is an exacting critic, and, most importantly, as befits a critic in the major paper of the industry town, he knows how to notice and assess details of production and give credit or blame where it is due.

This brings us to the newspaper of record, the *New York Times.* The *Times* is the leading exponent of a dictum that rears its head frequently in the world of television criticism: The British do it better. (I exempt from this, however, *Times* reporter and analyst of the television industry, Les Brown.) The *Times* critic who waves the Union Jack is John J. O'Connor, the paper's television critic for several years past. However, to do the dirty work of reviewing Hollywood programs, the aid of movie writer Tom Buckley has been enlisted.

At the *Times,* praise is given only grudgingly and is reserved primarily for heroes of the higher culture. Woody Allen is in, nearly all of American television is out. Since the *Times* is the paper of record, however, due notice must be paid. Buckley is assigned the chore of demolishing entertainment ephemera, which O'Connor also sometimes deplores at greater length in wordy Sunday think pieces. But, for the most part, O'Connor writes about British imports and PBS documentaries.

Come to think of it, the *Times*'s strategy seems to be to make television appear so boring that *Times* readers will lose interest in the medium. The idea that television might be fun, that it is popular culture with a range of content and value from the ridiculous to the sublime, is no more fit to print in the *Times* than would be Michael Arlen's notion that television watching has sexual connotations.

In a historic moment some years back, a daring delegation from the *Times*'s cultural department put on their tan raincoats and stepped into Times Square one lunch hour to view the X-rated movie, *Deep Throat*—to see what all the fuss was about, of course. One wants to say to the *Times* cultural department: "Put on your tan raincoats again and watch a little television."

I don't know if Tom Shales wears a tan raincoat, but he seems to look at a lot of television. Shales took over as chief television writer for the *Washington Post* when Sander Vanocur disappeared into the vaults at ABC News, and, lo and behold, a major American newspaper had a television critic who visibly gave off signs that he enjoyed television. That it mattered to him. That if it was lousy, it didn't mean that everyone in Hollywood and at the networks was a bonehead producing garbage, it meant that his fun was spoiled. But he could still have fun writing about how lousy it was.

Shales may be the first child of mass culture who grew up to be a television critic. He knows what a comic book is, a roller skate, a Rolling Stone. If Gilda Radner parodies Patti Smith on NBC's "Saturday Night Live" he doesn't have to poke a neighbor to ask what it's about. Unlike those television critics who find television an alien world, to Shales nothing seems to be alien, on television or off. It may be terrible, but it's not alien.

If Shales has any writer on whom he models himself, it may be that old curmudgeon mentioned earlier, H. L. Mencken. Shales has only one weakness Mencken rarely showed—a soft heart. After laying about at television with his quick wit for a time, Shales inexplicably relents and praises some program he has no business liking. Usually it's a PBS show, and probably pressure from PBS had been driving his bosses crazy. Yet, even with that flaw, Shales is hands down the most interesting television critic I know of who writes for daily newspapers.

I may have missed someone, of course. The writings of Rick DuBrow of the *Los Angeles Herald-Examiner* or Frank Swertlow of the *Chicago Sun-Times* don't find their way to my provincial corner of Manhattan. And I have ignored television's critics of television, as well as other possible geniuses toiling in the wilderness.

Those I have slighted are invited to send me their tear sheets or cassettes for future revisions of this essay. I will certainly include you, but, being a critic of critics, I do not promise to be kind.

Epilogue: When published in *Emmy* magazine, this essay did elicit two responses. One was from Tom Shales, who appreciated the praise, and wrote, "Thank you from the bottom of my fat-encrusted little heart." The other was from Frank Rich, who appreciated the praise of his movie reviewing but not the criticism of his television reviewing. He took issue with the notion that he writes put-downs of setups, that *Time* has a corporate editorial style, or that it has made a corporate editorial decision to deplore television. Interested readers may consult back issues of *Time* and form their own judgments.

15. Twelve Television Reviews

In September 1978, I began reviewing television programs for *The Chronicle Review,* a fortnightly section of *The Chronicle of Higher Education* (subsequently re-christened *Books & Arts,* a year later it became a separate publication). Writing two television reviews a month—instead of on a daily or weekly basis, as do many of the critics considered in "On Television Critics and Criticism"—has both advantages and drawbacks. The drawbacks include: a tendency to focus on "specials," thus failing to give an accurate picture of the normal flow of television programming, good, bad, and indifferent; and a similar predilection, perhaps, to find programs worthy of praise, with the idea that one's readers are wary of television or snide about it, and need to be coaxed to watch the "best of TV." The advantage, of course, is that one is exempt from the daily grind; this in particular provides an opportunity to reflect on the series programs, watch a number of episodes over a period of time, and review television in a manner more closely attuned to the way regular viewers develop a sense of continuity and familiarity with their favorite shows from week to week. The following twelve reviews are a selection from the 1978–79 television season.

Lou Grant

Lou Grant is a commercial television series that persons of intellect are not ashamed to admit they watch. It failed ignominiously in the ratings when it was launched last fall, but CBS, rather uncomfortably, stuck with it. The network was rewarded during the summer re-runs, when the show climbed one week to the head of the list. Recently its star, Ed Asner, and a supporting player, Nancy Marchand, won Emmys for their first-year performances. "Lou Grant" is taking its place among the small group of shows the networks call "quality television," and it's worth our attention to see what, in this case, "quality television" means.

The fictional Lou Grant is, of course, a survivor from "The Mary Tyler Moore Show," where he was the crusty but softhearted head of the newsroom at a Minneapolis television station. After the demise of that popular show (also a slow starter) he was transmogrified into the city editor of a Los Angeles newspaper for the new, hour-long program. The enormous success of the motion picture *All the President's Men* may have prompted MTM Enterprises, the production company, to revive the moribund newspaper genre.

"Lou Grant" is in fact so derivative of that movie that the writers make a virtue of the debt; they refer on occasion not only to Woodward and Bernstein but also to Hoffman and Redford. And with the self-reflexiveness that television, after years of reticence, is now adopting, Lou Grant even makes an oblique critique of his own prior existence. After talking on the telephone to a television news reporter, he says, with the slightest touch of irony, "They'll never get it right."

Lou Grant is indeed a different person in the newspaper city room from the man who was Mary Tyler Moore's boss and surrogate father. Asner plays him with less whimsy. He's a stronger character with a wider range of feelings. No longer merely a curmudgeon with a heart of gold, he can focus his anger, express moral outrage, take a stand on principle. The hard man with the soft interior has become a gentle man with a vein of iron.

Around Grant is built the ensemble acting that knits together a television series from episode to episode. The supporting parts in

"Lou Grant" include two young investigative reporters, Joe Rossi and Billie Newman, played, respectively, by Robert Walden and Linda Kelsey; the managing editor, Charlie Hume, played by Mason Adams; and the publisher, Mrs. Pynchon, portrayed by Nancy Marchand. In the programs I have seen, Rossi and Newman are most often the catalysts to move the plot along, while the older man and woman, Mr. Hume and Mrs. Pynchon, are more richly delineated, multidimensional characters.

This superior cast meshes well. But that alone does not constitute "quality television"; you can find good acting in some of the tritest shows on the air. "Quality television" is based rather in a sense that television entertainment need not be trivial: that it can be entertainment and yet address some of the social concerns of our time. (I am putting "quality television" in quotes because it refers to an attitude toward tone and subject matter; some shows—say, "The Carol Burnett Show"—can be television of very high quality without being "quality television.")

"Lou Grant" could have gone the route of many a departed and unlamented newspaper series and used the newsroom setting solely as a foundation for melodrama—reporters, like lawyers and police, are legitimate presences in the world of vice and violence that television action-adventure programs portray. Sometimes, indeed, "Lou Grant" slips into that mold. But the production team for the show wanted to attempt something more relevant to the actual world of newspapers and their role in American society. They wanted to create not just entertainment, but entertainment with the feel of significance: in short, "quality television."

The premiere episode this fall centered on the issue of a reporter's right to protect his sources, a timely subject because the program was broadcast on the heels of the Supreme Court decision permitting access to newsrooms by authorities with search warrants. The show aired, moreover, while the *New York Times* reporter Myron Farber was fighting a contempt citation for refusing to turn over his notes to a judge in a murder trial. The second episode dealt with the newspaper's responsibility to publicize incidents of torture in a Latin American dictatorship unmistakably similar to Nicaragua, and it was broadcast only a week or two after a major battle between rebel forces and troops loyal to the

Somoza regime. If the series were any more timely it would almost be news instead of entertainment.

But entertainment it emphatically is. These subjects are raised not as tokens for debate or topics to be elucidated, but within the framework of television's entertainment cannons. There has to be a story, and then a "backstory" to thicken the narrative brew, and above all a denouement after 52 minutes. More often than not, of course, there are no simple answers to the issues "Lou Grant" touches on—or its producers, like the rest of us, don't know them. But television entertainment doesn't work that way. Audiences can't be left confused, frustrated, in a cloud. One of the social functions of entertainment is to provide satisfying endings, a sense of order and understanding not so readily found in daily life.

But what, in the context of commercial television, is a satisfying ending? Certainly not something that takes a clear and definitive stand on controversial public issues. That might alientate some viewers, not to mention sponsors. "Quality television" creates an appearance of dealing with the issues of our time, but takes care to neutralize every clear stand taken by one or another character on a program. The result is to leave viewers with the impression that they have seen important contemporary themes seriously treated, while, in fact, the viewers come away no less confused, frustrated, or in a cloud than before.

Can "Lou Grant" take sides on whether a reporter's asserted right to protect his sources holds precedence over an accused person's right to confront his accusers? In that premiere episode, written by Michelle Gallery and directed by Jay Sandrich, a venal doctor prescribes drugs to kids in the guise of diet pills. One girl overdoses and dies. Her boyfriend, seeking evidence against the doctor, breaks into his office, steals his records, and gives them to reporter Rossi. "Don't tell me how you got them," Rossi says. "I can't know"—words almost immediately echoed by Lou Grant to Rossi.

One of the high points of this particular show is a superbly directed scene in the publisher's office as the publisher, her lawyer, the managing editor, and the city editor discuss the ethics of journalism. Potentially didactic writing is brought to life by fast-paced acting and camera work, and we are left with the clear impression that "principled" journalists are as hypocritical and opportunistic as those they want to expose, except that common morality is on the journalists' side.

But the law is on a different side, and Rossi goes to jail to protect his source. Hey, this is serious; maybe for once the modern equivalent of the U.S. Cavalry won't be riding to the rescue. But don't get your hopes up. The "source"—the youth who stole the files—turns himself in, out of admiration for Rossi. Meanwhile, in the "backstory," a bunch of kid reporters comes up with damning evidence against the doctor—evidence that's not tainted like the stolen files.

The episode about the Latin American dictatorship, written by Seth Freeman and directed by the show's executive producer, Gene Reynolds, takes much the same path. The dictator's wife is visiting Los Angeles, and, lo and behold, happens to be a friend and guest of the socialite publisher. Anti-dictatorship pickets protest the visit, but the managing editor inexplicably wants to play down the story. It turns out he's trying to repress his own memories of once having been among the dictator's victims. He insults the wife at a publisher's party, then decides to purge his nightmares by writing an exposé of the dictator's torture methods.

The question is, Should the exposé be published? Is it up-to-date news, is it responsible policy (alienating the dictator could mean the cancellation of a deal for fighter planes, one editor says, jeopardizing Los Angeles-area employment), is it good for the publisher's social life? A stand for human rights wins out over other considerations. The article is printed. The angry wife denies it all, but is shamed into meeting the protesters, and—in an ambiguous ending—it appears that she has seen the truth about her husband's regime. She praises the dissidents, which could mean that the regime may soften, or that she will incur her husband's murderous wrath.

"Quality television" is a magic act: Now you see social significance, now you don't. The art of writing popular entertainment is to create a structure that the casual viewer will accept as serious even while the serious themes are carefully balanced and hedged. It takes time and an inclination to ponder that structure to discover that the resulting even-handedness is not a clear presentation of opposing sides; it is an intellectual muddle. The only point to add is that the forms and aims of "quality television" are not unique to television; their origins can be traced a long way back in literature, theater, and movies.

The New Klan and A Question of Love

With its insatiable appetite for fresh material, American television relies heavily on actual events for subject matter. No sooner has there occurred a trial, a trend, or a tragedy than some enterprising producer is making a program about it—a documentary if it's for public television (which seems to be paying more attention to current events these days) and a fictionalized version, or docudrama, if it's for a commercial network, because documentaries do poorly in the ratings.

Two interesting examples of the translation of actual events to the small screen are an hour-long documentary, "The New Klan: Heritage of Hate," on public television stations, and on the ABC network, " A Question of Love," a teleplay based on a recent court case involving a father's effort to gain custody of a child living with a lesbian mother.

Both programs, as it happens, are concerned with the lives of working-class people in the South and Southwest. A legitimate question in both cases—but particularly in "The New Klan"—is how the fact of social class, or the perspective on class adopted by the respective producers, affects their interpretation of events and our perception of them.

"The New Klan" was produced by independent filmmakers Eleanor Bingham and Leslie Schatz, and financed by grants from the Corporation for Public Broadcasting and public-television stations. It takes a far broader approach than the title indicates: the program deals with the surviving remnants of the traditional Ku Klux Klan and also with the historical Klan through newsreel and archival footage dating back to the 1920s. In fact, to my taste, it deals with far too much. It races from shot to shot, detail to detail, as if facts alone could make up for the absence of a point of view.

Of course it's not entirely fair to say that the program lacks a point of view; what it lacks is an adequate explanation for the social phenomenon it presents to us. The view is that the men and women of the Klan are racists who hate blacks and Jews and who deserve to be condemned for their prejudice. Moreover, as the eye of the camera searches them out, these racists tend to be stupid and obese, grotesque characters who people the American landscape. They rise from among the rednecks of the American sunbelt, from their trailer parks and bowling alleys, and from fanatical Christianity gone sour.

But out of this heritage has come a handsome, smooth-talking young man, David Duke, still in his late twenties. Duke, the leader of the New Klan, has developed a following in Southern California and has become something of a media figure. His personality and philosophy provide the thread that holds "The New Klan" together, and to debunk him is one of the program's goals. To this end the producers use a taped interview between Duke and *New York Times* correspondent Wayne King, himself a southerner, in which King softly, doggedly exposes Duke's pretensions. His ace-in-the-hole, and the program's final image, is a photo showing Duke as a teenager in a Nazi uniform, holding a placard that reads, "Gas the Chicago 7." Even the cleverest of the Klansmen is no more than a rabid would-be Hitler.

An exposé mentality pervades "The New Klan," as if depicting the racism and intolerance of the Klan were enough to provide an insight into its place in American society. "Can it happen again?" the film portentously asks at the end, and the viewer has no way of answering, because the program treats the Klan phenomenon as if it were a fever or a spell of stormy weather, something we are powerless to anticipate or prevent.

The unfortunate thing is that the documentary contains the seeds of a more adequate explanation, but the filmmakers chose not to develop the material they possessed.

Ironically, the deeper theme is sounded in the film clearly and unmistakably in two brief appearances by black political figures. On the first occasion, a televised debate between Duke and the Rev. Jesse Jackson, it is Jackson who says that the real issue is not black character or behavior, but jobs. Later, Charles Evers is shown with two former Klan troublemakers, and Evers remarks that "the Klan had the answer to the poor white people's problems."

Is it that blacks can see that issues of class and employment are central to the Klan's appeal while Klan members themselves cannot? Apparently not. The filmmakers show a discussion among members of the New Klan in Southern California. "Attract new industry to San Diego?" asks one. "What for? Just to give it to Mexicans? If we could get the green-carders out and the illegals out there'd be plenty of jobs."

Cut immediately to a shot of a racist comic book. It is the filmmakers who glance away from the economic and social issues, not the protagonists themselves. The documentary is bereft of a social perspective. We never hear what jobs the Klansmen in fact hold; we don't learn whether they're unemployed or in competition with blacks and Chicanos for work.

Public television is no less susceptible than commercial television to giving us sensation in lieu of explanation. Filmmakers sometimes defend themselves by saying they deal in images, not explanations, as if images didn't have a point of view as well. But the producers of "The New Klan" can't use that one. Their voice-over narration is as heavy-handed as any "Voice of God" network news special. That omniscient voice simply didn't want to tell us all it knew.

The producers of "A Question of Love" have, as might be expected, a somewhat easier time coping with issues of class. Their central concerns are sexual preference and the family; their protagonists are working-class people because that's the way it was in the actual Texas court case from which the events of the docudrama, and even some of its courtroom testimony, are drawn.

Instead of actual working-class people, the fictionalized version gives us Gena Rowlands as the mother and Jane Alexander as her friend, two exceptional actresses who can appeal to our feelings or depict complex emotions far more effectively than the average person. Moreover, the program unabashedly supports its subjects, rather than seeking to put them down. Nevertheless, the docudrama has its ways of copping out just as does the documentary.

The problem in docudrama is to choose between fidelity to the actual event on which it is based and fidelity to dramatic tone and coherence. Often it's impossible to be true to both at once. In this case the script by William Blinn and the direction by Jerry Thorpe obviously tend to endorse the mother's claim to her child, and all our emotional energies are marshalled on her side—only to suffer a letdown at the end when the jury, true to life, awards custody to the father.

Gena Rowlands plays a nurse with two children, a teenage boy and a boy of nine. As the program opens they are moving in with the mother's lover, a bank employee, and her daughter. The older son notices a joint bank account and a double bed, and jumps to a conclusion. Later, while driving with his mother—a knowing touch, to put the conversation in an automobile where so many significant American conversations take place—he asks her if she's a lesbian, and she affirms it. Anxious over his own sexual identity in the eyes of his peers, he grows estranged from her, moves in with his father, and sets into action the custody fight for the younger brother.

The program, as the real-life case, is a test of the mother's fitness to raise her younger son. Although the actual mother lost the test, the actress wins it for her in this fictional depiction. In recent years Rowlands has brilliantly played histrionic and/or hysterical women in the films of her husband, John Cassavetes, notably in *A Woman Under the Influence* and *Opening Night*. Here she is more controlled, more intense by holding in her emotions, though there are moments when her rapid play of expression may verge on over-acting.

The boy in question unmistakably prefers to live with his mother. And the boy's maternal grandmother, after an initial violent revulsion against her daughter, comes to support her; "Gotta stand with your own," she says. If we are not convinced by the dignity and moral worth of the mother and her friend, the father, an airline mechanic, is portrayed by Clu Gulager as a slow-witted, hard-headed redneck of decidedly questionable character.

Though muted and never direct, the program clearly presents what may be called a feminist point of view. "A Question of

Love" tells us that true human feelings are more important than conventional prejudices. Here again, however, in following the line of actual events, the producers fall into an anomaly. The mother's lawyers, seeking to impugn the father's virtue, discover that he has impregnated a young woman and paid for her abortion. The anti-abortion tone of this material may have come from the actual court case, where it could have been tactically useful before a Texas jury, but it goes against the predominant grain of the show.

Mork & Mindy

The most popular new prime-time program of the 1978–79 television season is ABC's "Mork & Mindy," a situation comedy about a visitor from outer space who lands in Boulder, Colo. For the first seven weeks of the fall it ranked third in the overall ratings behind "Three's Company" and "Laverne & Shirley," and no doubt there are those who regard such news as confirmation of television's debasement—or infantilization—of the public taste. But "Mork & Mindy" is more than just another formula sit-com, and it is of interest beyond the narrow range of the Nielsen-ratings sweepstakes.

Think, to begin with, of the premise: A young man from the planet Ork is sent to Earth to observe and report back on the natives. Like an apprentice anthropologist doing his first field work, he is prey to all sorts of misapprehensions, false deductions, and crossed signals. None of the customs or usages of Earth people can be taken for granted; all of our conventional ways of talking and acting are called into question. The comic potential of such cognitive dissonance is obvious.

Then, of course, there is the other side of the coin—the language, styles, and powers of the intergalactic visitor. Besides his funny words and odd habits, like "sitting down" on his face instead of his rear end, Mork of Ork is not obliged to operate by many of the physical laws of Earth. He can play with time (though not space) as we cannot.

Supernatural powers are not new to television—remember "I Dream of Jeannie," not to speak of "Superman"—but "Mork &

Mindy" nevertheless represents a significant innovation in the present television scene.

Back in the 1950s, when George Reeves as Superman was leaping tall buildings in a single bound, television commercials tended to be stodgy, prosaic, literal, unimaginative: the prototypical "doctor" in a white coat holding a bottle of pills up to the camera, Betty Furness extolling the virtues of a refrigerator's ice-cube tray. All that has changed. Now it's the commercials that are most often exciting visually and technically, while entertainment programs, particularly situation comedies, are stuck in a rut of conventional realist techniques. When you see two or three characters talking at each other on a studio set, and then suddenly you're transported to the bottom of a cereal box, or McDonald's bombards you with a stream of one-second vignettes of human joy, it makes you wonder whether programs are deliberately kept tame and slow so that the commercials will be even more stimulating in comparison.

"Mork & Mindy" is the first situation comedy in some time to avail itself of the magical techniques that are now common in commercials. Moreover, the program uses advanced video technology (although in cautious moderation) to produce special effects that previously have been seen primarily in video art.

None of these technical opportunities would matter, however, without the critical element, a genuine comic personality to exploit them. It looks like "Mork & Mindy" has its man in a promising young comic talent named Robin Williams. It's always risky to make premature claims for young actors on commercial television, particularly for comedians. The rigid formulas and repetitive banalities of network sit-coms can turn an interesting fresh face into a predictable caricature within a year or two—witness John Ritter on "Three's Company" or Henry Winkler, the "Fonz" of "Happy Days." So far, however, Williams is new enough to enjoy—and too new for us to recognize his limits.

In repose, he is difficult to describe—a small man with longish dark hair, bright blue mischievous eyes, prominent nose and chin, and a malleable mouth. But then he is rarely in repose. Williams is an accomplished mimic and verbal impersonator. His vocal skills seem ready-made for situation comedy, with its well-known emphasis on verbal rather than visual humor. He is a manic fount of noises and fast-changing voices. And he has a mobile face that can assume the varied personae of his many voices.

It will take a while to figure out how much of Mork is Robin Williams and how much is the brainchild of the show's writers and producers. What seems most likely is that they found Williams and built a concept and a character around his special talents.

Williams was scouted at the Los Angeles showcase nightclubs for new comedians (the Comedy Store, the Improvisation) and given a guest appearance on "Happy Days" as Mork from Ork. When that went over well, the new show was put into development by the same production unit at Paramount that produces "Laverne & Shirley" and "Happy Days."

What the producers first saw at the Los Angeles showcase clubs was revealed to subscribers to Home Box Office, the cable pay-TV network, when it presented an hour-long Robin Williams special. The program, however, was taped at the more prestigious Roxy club in Los Angeles after "Mork & Mindy" had become an instant hit, so it showed not Williams the struggling young comic but Williams the overnight star. A good deal of his humor had to do with his apparent ambivalence over success and the price one pays for it in television.

Much of the program's humor also afforded Home Box Office the opportunity to show what *it* can do without the restraints of network censorship. Williams indulged in off-color language and gestures from start to finish (I hesitate to use the word "obscene" because of its legal implications). "Dirty" words have the shock power to make us laugh, but they can also distract from what's genuinely original in a comic's performance. Overall, however, the show had a loose, rough, free-floating atmosphere reminiscent of the early slapdash days of television, refreshing in contrast to the conventional, closed style of network programming.

Williams's most brilliant turn was what might be called a subjective correlative to his inner state of mind. He pantomimed opening up his head to reveal what goes on inside his brain. In an incredible sequence so rapid-fire I couldn't begin fully to describe it after only one viewing, he then performed like an engineer inside the control room of his brain, swiftly inserting programmed instructions on who to be or what to do, trying to keep a balance between the ego and superego struggling for hegemony, while holding off the dreaded id, which burst through to expression in critical moments. Williams's mercurial skills were on display as he changed voices and facial expressions with dazzling quickness.

Several times during the special, members of the audience called out for Williams to perform as Mork. But Williams repeatedly rejected the suggestions, sometimes employing clever ad-libs, once stating directly, "No, I have to do that five days a week. This is a special time."

Whether or not Mork can contain the best of Robin Williams remains to be seen. On the series, he has had an opportunity to perform many of his voices, including the popular parody of a Shakespearean actor he uses in his nightclub act (Williams had done a bit of serious Shakespearean acting early in his career).

Ultimately, I suspect, the success of "Mork & Mindy" will depend on whether the show's writers can create situations that extend Williams's talent, rather than restraining it in the strait-jacket of a formula. Several of the episodes so far have been pedestrian enough to caution against getting one's hopes up.

One episode, however, has shown off Williams and "Mork & Mindy" at the top of their form. It was aired, curiously, on Thanksgiving evening and thus fell far down in the weekly Nielsens. A bully is hassling Mork's friend Mindy (Pam Dawber, who plays straight woman to Williams's Mork) in a restaurant. Mork's wacky literalism in response to the bully's demands—when the bully says "take that back," Mork repeats all his previous actions in reverse—confuses and outwits the tough guy, who threatens revenge.

There must be a showdown between Mork and the villain; but Orkans, it turns out, are nonviolent. On Ork it's humiliating to fight, but Mork discovers that his values have no currency on Earth, where he is seen only as a coward. Mork will not fight the bully with his fists, but when he's knocked down he decides to retaliate with his supernatural powers. "I'm into time warps," he says, and puts the bully into slow motion. This makes for a splendid sight gag as the bully (played by Brion James) throws his punches with excruciating slowness while Mork dances around him and finally gives him a nudge that topples him over.

At the end of each half-hour Mork reports back to Orson, his Orkan superior, on what he has learned about Earth.

"I'm trying to fit in on Earth," he says, "but these are violent, violent people." Then he reels off some of the violent things we do—like "slashing" prices and "crashing" at a friend's place. The show uses both verbal and visual humor in a lighthearted way to make a valid statement about human behavior. That's not infantilization; that's, to coin a phrase, "adultilization" of the public taste.

We Interrupt This Week

This article started out to be a post-mortem on a public-television series that relatively few people have had an opportunity to see. "We Interrupt This Week," billed as the intellectual's game show, appeared last fall on fewer than three dozen of the nearly 300 public-television stations. Although it attracted considerable attention in the press and some public notice in New York (where it is produced) and other cities where it has played, the show came very close to being dropped when its initial financing ran out early this month.

But, at the last minute, the Corporation for Public Broadcasting—responding to a letter-writing campaign by several thousand of the program's fans—came up with an extra $100,000 to give the series five more weeks of life, while further financing is sought.

"We Interrupt This Week" is the first game show produced for public television, and its promoters harked back for antecedents to "Information Please!" and "Quiz Kids" and other intellectually strenuous entertainment programs from the golden age of radio. (No one, so far as I know, thought to appeal to contemporary game-show enthusiasts by calling it "Manhattan Squares.")

More apt comparisons were made to the famous 1960's British series satirizing contemporary events, "That Was the Week That

Was" ("TW3"), and to a later BBC show on current events, "Quiz of the Week." The creator and producer of "TW3" and host of "Quiz of the Week" was a dapper actor and entertainer, Ned Sherrin, who came to New York last year to direct and perform in a Broadway musical, "Side by Side by Sondheim." When that production closed, Sherrin sold the idea of a game show, with himself as host, to the New York public-television station, WNET. The Corporation for Public Broadcasting and those three dozen or so public-television stations also put up some money, and Sherrin's new show was launched.

"We Interrupt This Week" thus became another example of public television's unquenchable passion for things British. A passion, to be fair, shared more moderately by the commercial American networks: such successful situation comedies as "All in the Family," "Sanford & Son," and even the currently popular "Three's Company" were based directly on British counterparts. Where the commercial programs merely have borrowed British ideas, however, "We Interrupt This Week" centers on a British personality.

Sherrin opens each program with a series of one-liners on the week's events. Then he begins firing questions at two competing panels of "extraordinarily erudite experts," as he calls them. One is a "home" team of three regular panelists, the other a trio of guests. Points are awarded for correct answers and, at Sherrin's whim, for clever ripostes.

"My decisions," he announces each week, "will be arbitrary, prejudicial, and final."

Precisely. This skillful British actor is playing not merely himself, or appearing as a British television personality; he is portraying the American fantasy of the urbane British intellectual—quick-witted, sharp-tongued, imperturbable, and ineffably superior. He performs not only as host and quizmaster but as our image of the British schoolmaster. Some of the panelists respond by approaching the experience as an oral exam.

The show is at its best in those all-too-rare moments when one of the American panelists musters the presence of mind to match wits with the Briton. Sherrin, it must be said, holds the upper hand (not only by being British) because his monologues and all the questions are prepared for him by writers Tony Geiss and Harvey Jacobs.

Author and journalist Jeff Greenfield, one of the regulars, is the funniest panelist I have seen on the show. No doubt he gains courage by his familiarity with Sherrin, but of the Americans he seems the least cowed by trans-Atlantic competition. He presents a persona of a street-smart New York wise guy, one of the few American types immune to Anglophile angst.

Besides the occasional *bon mot,* what rewards does an "intellectual's game show" offer to intelligent viewers? There are no prizes to win, no costumed competitors to root for (or against) vicariously, no embarrassing revelations to savor; the brain game is not "The Dating Game." Nevertheless, the premise of a highbrow game show differs little, I think, from the most lowbrow; the spirit of competition animates both. Either you get involved in the struggle between the program's teams, or you pit your own wits against those of the on-screen competitors. "We Interrupt This Week" fails to accommodate either of these competitive urges, and therein lays its most glaring weakness.

I came to realize this when I spent several afternoons observing the show's taping sessions at WNET's Studio 55 in New York. The studio setting provided a completely different impression of the program from that I had got off the screen. The competitors were seated at two tables in a semicircle with Sherrin facing them in the center. An audience of abut 100 sat in tiers behind the contestants, and the spectators seemed naturally to take sides based on the happenstance of which team they were sitting behind. The "live" audience witnessed the entire proceedings in a single field of vision, and it cheered and applauded the game enthusiastically.

Most of this experience is lost in translation to the screen. The problems lie both in set design and in the direction. Watching at home, you never get a sense of the whole; no camera is placed to take in all six competitors, though occasionally there are split shots in which one team appears above the other on the screen.

There are too many isolated shots of individuals, when the interaction among teammates and between teams gives the show its flavor. And when individuals are shown, the camera angle cuts off their nameplates, so that viewers who tune in late don't know who the contestants are. The director is sluggish in changing shots, often missing on-the-screen gestures and expressions that the studio audience is enjoying. And there are far, far too many single shots of Sherrin.

Some of these problems are inevitable in "breaking in" a show, particularly in public television, where budgets and facilities may be limited. "The Dick Cavett Show" had shakedown difficulties on PBS at least as severe as those on "We Interrupt This Week." (The Cavett show changed directors and sets en route to its present, more satisfying format.) With "We Interrupt This Week," unfortunately, the problems go beyond the technical to the conceptual.

The show was designed as a quiz on the news of the week. The questions written for Sherrin essentially are riddles on current events. This format could provide a bracing challenge to the viewer, but—as it has worked out—many in the home audience are denied the pleasure of competing against the panelists: All too many questions seem designed only for compulsive readers of the back pages of the *New York Times* and the *Washington Post.* And Sherrin speeds from question to question so rapidly that the possibility for panelists to build on each other's remarks, playing little word games within the larger game, is continually aborted.

"We Interrupt This Week," normally a half-hour show, began 1979 with a one-hour special, "We Interrupt This Year: 1978," carried to all PBS stations on a Sunday evening. Perhaps because of the expanded time frame, or perhaps because the show was at last reaching the public-television masses, the year-end special had a much different feel from the regular programs.

The pace was less hectic; there were fewer and simpler questions, taken more from the front pages than the back. The director found more to look at on the set than repeated head shots of host Sherrin. The panelists had more time to say witty things to each other, and the director sometimes caught them at it.

This show (at the time of taping, it was believed to be the swan song of the series) pointed to a more effective style for the intellectual's game show—if it should survive beyond its brief reprieve.*

* It didn't.

Roots: The Next Generations

The most popular event in the history of American television—perhaps in the history of American mass culture—returns this month for an encore. Two years after "Roots" drew the largest television audience of all time, the series based on Alex Haley's book continues on ABC stations with seven two-hour episodes under the general title of "Roots: The Next Generations"—the saga of Haley and his ancestors up to the present.

The original "Roots" was a striking media and social event. According to the Nielsen ratings, some 85 percent of the American households that have television sets tuned in to at least part of the 12-hour series. This enormous majority of the American people saw what television entertainment had shown rarely, if ever, before—admirable, heroic black people struggling to survive the degradations of slavery; black men and women raising families, loving and grieving, and living the human story from generation to generation.

Despite its acclaim, its smashing of taboos, its compelling appeal to millions of people regardless of color, "Roots" made surprisingly little impact on the television medium. The presentation of blacks on network programs has changed little, if at all, in the past two years. And one suspects that the same might be said for attitudes and perceptions in the wider social context. Nonetheless, "Roots" was a remarkable educational event. Although not the first television series to serve as a subject for credit courses at colleges and universities, it greatly expanded the use of dramatic television for pedagogical purposes. Many thousands more people were exposed to information about (and an interpretation of) American slavery than had ever studied the subject in a college history course. "Roots" also brought to the fore questions about the translation of history into dramatic entertainment.

David L. Wolper, the executive producer, though armed with a yard-high stack of reviews in praise of "Roots," was particularly miffed at a small minority of critics, some of them professors who took issue with the historical accuracy or interpretation presented in the series. He insisted that the demands of dramatic construction took precedence over all else; let the professors try to tell the story of slavery *their* way to a popular audience, he said with a certain disdain, and see if anyone would watch.

There is no simple way to solve the dilemmas posed by competing standards and conventions of scholarship and of mass entertainment. But if commercial (or public) television would deal with serious historical subjects more often than on rare special occasions like "Roots," we might have a greater variety of models for presenting popular history.

The Wolper model descends from the 19th-century theatrical melodrama; those critics were perceptive who found the most apt comparison for "Roots" in Harriet Beecher Stowe's novel, *Uncle Tom's Cabin,* which played for decades as a stage melodrama. Good is good and evil is evil, and rarely are they allowed to mingle in one human character. Family values are central to the drama; patriarchy is taken as the norm, but women play significant roles as moral touchstones, as activators of romance, as bearers of the important principle of love. To borrow Stephen Crane's description of melodrama, "Roots" and "Roots: The Next Generations" are works of "transcendental realism."

The simplicity and clarity of the characterizations are carried over into the style. The lighting is bright and flat, the language repetitive of key words and phrases, the plot a solid frame of family history and generational change within which the wider themes of black history are presented. A young couple (who are to become Alex Haley's parents) kiss to seal their love; the woman looks up, changes expression; and we see what she sees: a Ku Klux Klan cross burning on a distant hillside.

Yes, as Wolper has planned it, the values of love and family bind our emotions to these black protagonists, and allow us to experience emotionally their fate as Americans. Wolper is able to have it both ways: To critics of the historical interpretation in "Roots" he replied that the events depicted were not meant to represent the lives of all blacks; they were true simply for Alex Haley's family.

The events of "Roots: The Next Generations" begin in 1882 in Henning, Tennessee, Alex Haley's boyhood home. The first three episodes carry the story forward to 1917, the year of America's entry into World War I, when Alex Haley's parents are still courting. Later episodes deal with Haley's upbringing, his career as a journalist (James Earl Jones plays the mature Haley), and the search he undertook to discover his ancestry that led to the writing of *Roots*.

Woven into the framework of family history are major issues of black life: the post-Reconstruction shift of black voters from the Republican to the Democratic party and their rapid disfranchisement; the rise of Jim Crow laws, sharecropping, the convict lease system, and lynching. Haley's maternal grandfather, Will Palmer, becomes the owner of a lumber yard in Henning, and black capitalism becomes a theme.

Wolper's ideology is conventionally American: strong as the ties of family are, they are rarely allowed to hold back the young. The goal is to "go forward," to strive for individual improvement, to serve through leadership. What stands in the way of blacks is white racism in all its forms, from subtle paternalism to murderous hatred.

It was often said that "Roots" succeeded because the series gave to blacks roles as parents, lovers, men and women of courage, that had been theretofore, in the popular media, reserved exclusively for whites. Less frequently was it pointed out that "Roots" not only gave to blacks, it took away from whites: Blacks were the heroes, whites the villains of this turnabout melodrama. It is in this regard that Wolper has made his one substantial change in the successor to "Roots." Now an occasional weak, supercilious, or even venal black character appears—although no black can yet commit a sin that compares with the cruelty and viciousness of the whites. And although most of the white characters remain evil, a few in the new series do escape the stigma of the "master race."

There is Jim Warner, the gentle poet and scion of plantation aristocracy, who falls in love with and marries a black schoolteacher, and finds that his family regards him evermore as a "nigger." There is the philanthropic liberal who pays college tuition for Alex Haley's father. Most significant of all is Goldstein the Jewish tailor, who treats blacks as equals and whose shop is burned by the Klan.

"Your Klan is pikers," he nevertheless tells the blacks, and describes the raging violence of a Russian pogrom. Wide-eyed, the blacks discover that fair-skinned people are also persecuted.

The "transcendental realism" of "Roots: The Next Generations" will sometimes put a lump in your throat or anger in your heart. But the question remains: What good is melodrama as history?

In the second episode we see the implacable force of white villainy cheat an unfortunate black man, send him unjustly to prison, whip him unmercifully, hunt him down, and hang him from a tree. The final shots of the lynching are presented as through the victim's eyes. Slowly the camera pans the circle of white faces, curious, ugly, impassive, watching a black man die. It is an affecting sequence, but ultimately an empty one. Making us feel like victims in this manner is sensationalism without insight.

On the evidence of "Roots: The Next Generations," melodrama does not succeed as history. History is, or ought to be, an account of the complex nature of past lives and events. Of melodrama's several virtues, complexity is not one. The series becomes interesting as black history, or as American history, only when it takes time out from melodrama.

For the third installment, Ernest Kinoy, a leading television dramatist who wrote the first three episodes and supervised the others, has created one such sequence. In it, he depicts Simon Haley's summer job as a Pullman sleeping-car porter. Entirely through dramatic action, the writer (and director John Erman) reveals the hard lives of anonymous black men who work 24-hour days for poor wages and meager tips.

The blacks in this sequence are not all saintly. Some are portrayed as dissolute gamblers, one as a spy for the Pullman Company who finks on men who are tying to form a union. Though no names, dates, or organizations are mentioned, we learn about an incipient black labor movement. Simon's friend and mentor is fired for "talking union," and his parting words are, "I'm bound to organize."

This particular sequence reveals a dimension of the black social and political experience that the "Roots" vision of progress, with its emphasis on individual self-improvement and its penchant for the melodramatic, rarely touches upon.

Family

Kate is "the Avis of mothers," my friend says. "She tries harder." Kate is Kate Lawrence, the fictional mother who is the conscience and cohesive force of "Family," a dramatic series on ABC. As I talk to friends and colleagues about television, I find no other series currently on the commercial networks that commands more affection and loyalty in academic communities than "Family." But among the general public its ratings have been slipping, and the steps being taken to increase its popular appeal do not bode well for the continuing quality of this unique show.

"Family" is unique because it treats as drama the intimate relationships, events, and crises of quotidian family life in contemporary America. The astonishing fact is that no other prime-time evening program deals with private life, the way we live now, except as formula comedy.

Afternoon television, of course, is quite another matter. It's jampacked with the family rivalries, love affairs, broken marriages, career problems, drug and alcohol abuse, and growing-up pains (only a partial list) that "Family" brings to the evening airwaves. When "Family" first appeared, in fact, it was spoken of as a leading exemplar of a trend toward soap-opera subject matter during the prime-time hours.

What immediately distinguished "Family" from the daytime soaps, however, was its quality as drama. Instead of the interminable, rambling, convoluted, lugubrious pace of a soap-opera story, "Family" has strived for a tight dramatic coherence. It has provided aesthetic satisfaction by creating emotional conflict and resolving it within an hour-long frame.

The "Family" formula generally involves bringing in an outsider—troubled friend, former lover—to serve as catalyst for dramatic upheaval. This gives the show continued novelty while retaining a primary focus on the subtle gradations of change in feelings and understanding among the permanent cast, the family itself.

The Lawrences are no typical American family. In their comfortable Pasadena home with a detached guest house, where grown-up son Willie currently resides, they live the lives of an educated and affluent elite—which may be one reason they have caught on more solidly with academics than with the average viewer. Another is the sensitivity (as opposed to network television's more common verbal banter and visual slapstick) with which the Lawrences regard each other's feelings. Although it has an array of soap-opera-style problems, "Family" nevertheless manages to convey more of the positive side of family life, love, concern, and commitment than can be found almost anywhere else on television—morning, noon, or night.

Film and stage director Mike Nichols was involved in the development of "Family" in collaboration with Jay Presson Allen and producers Aaron Spelling and Leonard Goldberg. Spelling-Goldberg Productions is better known for its two other shows currently on network television, the action-adventure series "Charlie's Angels" and "Starsky and Hutch." To counteract the drop in ratings of "Family," Spelling-Goldberg began to inject into it familiar formulas from those successful crime programs. It's the kind of medicine all too likely to kill the patient.

The high quality of "Family" often has derived (as with much good drama) from the show's capacity to construct personal situations that reverberate with general or even universal meaning. The conflicts between parent and child, husband and wife, are recognizable in our own experience, or they ring true to our sense of what such an experience may be like. Crime shows like "Charlie's Angels" and "Starsky and Hutch" are apt to do just the opposite. They take some larger social problem and reduce it to the level of a personal situation—here is street crime, dope dealing, child molesting, suddenly forcing its terrors upon you. For experience, they substitute ideology.

A recent episode demonstrated the unfortunate effects of turning "Family" away from private experience toward public ideology. The outsider on this particular program was that common television crime cliché, the vengeful psychopath. He is out to get even with Doug Lawrence, the head of the "Family" household, who as a lawyer had represented the man's wife in divorce proceedings. (The fellow went to jail for assaulting his wife, but what role Doug had played in his imprisonment is not made clear.)

The main plot of the program concerns the villain's rising tide of terror against the Lawrence family—telephone calls in the night, a vandalized garden, threats painted on the garage door, ominous gestures toward the children. All of this, of course, projects as much real feeling as a cement block. It requires of James Broderick, who plays Doug, little more than brow-knitting, and of Paul Shenar, who plays the villain, only a prolonged smirk.

At last the time comes when Doug must take a stand. "We have to take care of ourselves," he says, and gets his old World War II revolver out of the closet. Kate, the conscience, protests. Guns are for killing, she says; there are children in the house. Kate, played by Sada Thompson, is normally the moral arbiter of "Family," but this time she is overruled. "Extreme situations call for extreme measures," Doug insists, heading off for target practice.

The moment arrives when the villain invades the house in dark of night and destroys the family photo album. Doug levels the gun at him, but the contemptuous bad guy merely smirks on, confident that good liberal Doug would never pull the trigger. Then he threatens harm to the teen-age daughter, Buddy. That is one step over the line. Doug shoots a bottle out of the astonished villain's hand.

"I'm a reasonable man," he says. "But you've pushed me beyond reason. I will protect the people I love by whatever means I can."

This ringing declaration belongs to the attitude Clint Eastwood popularized a few years ago in movies like *Dirty Harry:* violence must be fought by violence, the citizen must take the law into his own hands. Gentle Doug Lawrence takes up arms in service to a desperate idea of rescuing "Family" from cancellation by resorting to an obsolescent mass-entertainment fantasy.

Meanwhile, this "Family" episode had, as do most such television shows, a "back-story." The sub-plot involves the teenage daughter, Buddy, played by Kristy McNichol, in a crisis stemming from, of all things, a school math test.

Her best friend has cheated, borrowing answers from Buddy's exam. The teacher notes the similarity of the two exams, and accuses Buddy, not the friend, of cheating. On principle, Buddy will not tell on her friend, but the friend will not admit her cheating, either.

This marks a crisis in friendship. The friend confides in Buddy that she hadn't studied because of problems at home, but she doesn't want to incur her mother's wrath by confessing that family difficulties interfered with her school work. The friend suddenly reveals her resentment at Buddy's stable home life. She prefers to let Buddy suffer than to get in trouble with her mother.

In the end, however, friendship wins out. Buddy's friend has learned not to project the problems of one relationship into another. "You're my best friend," she tells Buddy. "I won't let what's between me and my mom affect me and you."

Here is what "Family" can do as can no other current show—build drama out of anger or jealousy beneath the surface of relationships, out of the private motives for public acts, the necessity for choice in values and behavior, the opportunities for self-discovery and growth. Could the producers of "Family" have made an hour-long episode out of a school math test? They and their writers and directors surely know how, and it would have been a great deal better than what they gave us instead.

Given the power of ratings and television economics, a show about private relations may not be able to survive for very long on prime-time television. But we'll never know that for a fact if "Family" is first destroyed by being re-made into a drawing-room "Starsky and Hutch."

Too Far To Go

The images are lush: a riot of greenery, trees, bushes, shrubs. It is a glen, or a dell, in a forest. No, it is a yard. The camera pulls back to reveal a lawn and a driveway and two children playfully washing a station wagon. The perspective widens and we see a spacious white frame house with a wide veranda and green shutters. A man and a woman appear. They are dressed a little too nattily in slightly passé styles—he in a cotton seersucker suit, she in a gray suit with a white blouse. They are drinking, and discussing how to tell the children that their marriage has ended.

This is the opening sequence of "To Far To Go," a made-for-television movie on NBC stations. The man and woman are named Richard and Joan Maple—the leafy imagery is no coincidence—and they are characters in short stories by John Updike that have been appearing in *The New Yorker* and other magazines since 1956.

"Too Far To Go" is the first significant adaptation from American literature to appear on network television in some time. Except for best sellers that can be turned into mammoth multi-part series, the networks have given short shrift to our national literature as a source for programming; and the same is true for public television.

Robert Geller, the executive producer responsible for "Too Far To Go," already has a record of accomplishment in bringing American fiction to television. He produced the acclaimed series of short films, "The American Short Story," on PBS. For this next step forward—a feature-length film on commercial television, adapted not from one story but from many—Geller engaged a distinguished production group. The playwright William Hanley wrote the teleplay, Fielder Cook directed, Elizabeth Swados (of "Runaways" and "Nightclub Cantata") composed the music.

The original conception, to be sure, remains Updike's. Viewers may recognize lines of dialogue they read half a lifetime ago in his short stories. Many of us have grown up with the Maples, married with them, raised children with them, divorced with them.

And this brilliantly designed production creates its visual effects as carefully as does Updike, the noted stylist, in his prose. Nothing is by chance in the decor created by art director Leon Munier and costume designer Joe Aulisi.

Take the clothes Dick and Jane Maple are wearing in the opening sequence. Two forty-ish adults in the late 1970s, they look dressed for the 1950s. And that of course is what they are, children of the 1950s, refugees from the 1950s, marked by that era's failures, redeemed by its strengths.

Updike has been chronicling this generation of men and women for a generation in time. One may quarrel with his view—once more overt in his work—that, in the absence of a teleology, men and women of the mid-20th century searched for design and purpose for their lives mainly by sexual adventuring. That, however, is in large part the story of the Maples. They can reach out to almost any member of the opposite sex except each other.

Two exceptionally talented actors, Michael Moriarty and Blythe Danner, play the Maples. These unflamboyant, versatile performers (they must depict their characters in early as well as middle adulthood) create, for us to see, depths of feeling that the couple cannot perceive in one another.

The Maples talk cleverly, but they seldom communicate. We see their lives more through his eyes than through hers. Dick Maple, an architect, is a man who needs to build ramparts of wit to hold back his anxiety and his jealousy. Jane Maple is a mystery to her husband and perhaps to herself; each recognizes her vulnerability, but neither knows how to cope with it—except, of course, through extramarital sex.

The story of their twenty-year marriage is told mainly in flashbacks from that opening moment when they have reached a decision to divorce. One scene shows Jane Maple sitting on the floor in a scarlet leotard, doing yoga exercises while Dick is questioning her about her sexual affairs. In the midst of her yoga postures she blithely tells him everything—only this scene, as it turns out, is not a flashback; it is Dick Maple's dream. His wife never appears so strong in life as in his unconscious.

He learns about her love affairs under entirely different circumstances. He begins to question her as they are dressing for a formal party; the children, their antennae picking up some special signals in the air, burst into the room with their own concocted problems, as if to avert by diverting. He continues questioning her in the car. There is no blithe spirit here, only pain and resignation . . . and, for him, curiosity. "It's like meeting you all over again," he says.

The well-laid plans to tell the children of their decision to divorce go for naught. With the whole family reunited at the dinner table (two older children, away on trips, have returned), Dick Maple suddenly breaks down in tears. This is another carefully crafted sequence, much of it shot from the father's point of view. The camera's angle, however, seems to be slightly lower than eye level, so that his view of wife and four children is dominated in the foreground by a mahogany salad bowl and a wicker bread basket. That 1950s generation, they learned to do everything with style except live.

Mother tells the children about the separation. The older son responds in anger and resentment. "We're just the little things you had," he says.

"You're not," protests the father, recovered from his tears. "You're the whole point."

Life did have a purpose for the Maples after all: to reproduce, to perpetuate the species, to "raise a family." Perhaps the public ideologies of the postwar era had as much to do, after all, with nature's design as with neo-capitalist consumption. Marry and breed. Nature's ends were served, but the romantic part of the ideology lingered on like fallout to contaminate the Maples' lives. They kept searching among others' spouses for the love they could not find with their own, when perhaps the only honest love they were meant to feel was for their children.

Moriarty and Danner accomplish the difficult feat of making us see the Maples' weaknesses without allowing us to feel any condescension toward them. Think of Michael Murphy's role as the husband in *An Unmarried Woman* or Diane Keaton's performance as the poet-daughter in *Interiors* and you will get an idea of the distinction I am trying to draw. "Too Far To Go" is a work that deserves to be evaluated in the company of these films and such others as *Annie Hall*.

The argument has been made that the recent films of Woody Allen and Paul Mazursky (along with films like Claudia Weill's *Girlfriends*) have been commercially successful because television has accustomed audiences to accept intimacy as well as spectacle on the screen. Whether or not this is the case (I doubt it), few television programs have probed beneath the surface of human feelings and relationships as deeply as does "Too Far To Go."

The title, by the way, comes from an episode in the Maples' lives when they are vacationing in Puerto Rico. Dick has been felled by stomach pains, and they return to their twin-bedded room in a hotel. As he lies on his bed recuperating from his illness, she sits on the other bed, and they exchange their usual venom-filled *bon mots*.

Facing the void of their feelings toward each other, Jane suddenly asks, "Do you want to come over?"

"I want to," says the supine husband, "but I can't. It's too far to go."

The Scarlet Letter

The Public Broadcasting Service is presenting its first significant challenge to British hegemony in the production of literary adaptations for the small screen. On consecutive evenings public-television stations will broadcast a dramatization of Nathaniel Hawthorne's classic American novel, *The Scarlet Letter,* in four hour-long episodes, created by an American cast and production crew.

Competition with the British—whose dramatic programs have been permitted for years to dominate public-television's prime-time hours—is clearly the intention of PBS. A PBS press release contends that this production brings "to public-television viewers programming that matches, if not exceeds, the quality and attention to detail characteristic of 'Upstairs / Downstairs' and other trans-Atlantic imports."

In matters of television (or of anything else) I am certainly no Anglophile. So it gives me no pleasure to express the opinion that this American production is a disappointment, and that it can hardly stand comparison with British literary adaptations.

The problem of "The Scarlet Letter"—I regard the production in its entirety as problematic—illuminates how the entire British approach is different from any that Americans have yet undertaken. Both the BBC and the British commercial companies regard literary adaptations for television primarily as dramatic entertainment. The stress is on both words—"drama" and "entertainment." The product is intended to be *television,* not pedagogy or transliterated literature.

The British can also count, I think, on an audience for whom the national literature is more comfortably a part of its common culture. I am not trying to exaggerate the literary literacy of the British viewing public. The vast majority of Britons undoubtedly prefers "The Rockford Files" and "Laverne & Shirley" to dramatizations of classic novels. I am suggesting that the "elite" audience in Britain is likely to be more conversant with the major works of its literary heritage than is the "elite" audience in the United States.

In all likelihood, the classic works of American literature, particularly those by the 19th-century masters, belong to the cultural repertory of only a small minority of educated Americans. This, of course, is a good reason why PBS should seek to bring these works to a wider audience. But the choice of *The Scarlet Letter* as the premiere work in an effort to develop literary adaptations for television seems an ill-advised choice. Among Hawthorne's works, *The House of the Seven Gables* or even *The Blithedale Romance* would have worked better.

The Scarlet Letter presents a number of difficulties for adapters. It speaks not just of one distant cultural epoch but of two: the New England of Hawthorne's own day (the mid-19th century) and the New England of the early 17th-century Puritan founders. Puritanism is an especially complex concept in the history of American culture, heavily freighted with emotional import, yet in its own historical reality, barely understood. And primitive Boston, hardly a dozen years old in 1642 when the novel's events begin, conjures up few images in the landscape of the American imagination.

Moreover, the subject matter creates additional hurdles for contemporary audiences. On the one hand, the very idea of the scarlet "A" on Hester Prynne's breast may appear anomalous to today's permissive society. On the other, Hawthorne's treatment of the Puritan obsession with Satan may seem all too familiar to the present age. Whether or not Hester Prynne's daughter Pearl is a demon child is of compelling concern to the Puritan community, and the theme is an integral part of the novel's rich moral complexity; but contemporary popular entertainment is already oversupplied with Satan's offspring, as in *Rosemary's Baby* and *The Omen.*

It seems clear that anyone who wishes to grapple with these difficulties needs a firm stylistic viewpoint of his or her own to give the audience a grip on the material. That was the case with the memorable (through now rarely seen) 1926 silent-film version of the novel, made in Hollywood for MGM by the Swedish director Victor Seastrom, with Lillian Gish as Hester and Lars Hanson, another Swede, as the transgressing minister, Arthur Dimmesdale.

In the 1920s, Puritanism was a symbol of sexual repression and oppressive community control of the individual, and Seastrom added to that common notion his own Scandinavian experience of a puritanical culture. The community was depicted as hypocritical and malevolent, subjecting Hester to cruel humiliations. Gish brought to her role the controlled passion that was characteristic of the best silent-screen acting, and the cinematographer, Hendrick Sartov, swathed her in symbolic shadows and light. The result was a masterpiece of the American silent cinema.

What public television's "The Scarlet Letter" most obviously lacks is its own stylistic vision. Rich Hauser, who produced and directed the adaptation for WGBH-Boston, seems to have opted for an uninspired literalism in transferring Hawthorne to the screen. The novel has 24 chapters; Part 1 of the adaptation dutifully covers Chapter 1 through 6; Part 2, Chapters 7 through 12, and so on.

The phrase "attention to detail" in the PBS press release appears even more ominous in this perspective. As with the Children's Television Workshop production, "The Best of Families," the quest for perfection in the details of historical minutiae ultimately overwhelms the purpose of dramatic entertainment. A set duplicating Boston of the 1640s was constructed for the production, yet the setting provides no sense of place, has no character of its own.

Detail can be a tricky business. At one point in the novel, during Dimmesdale's travail, Hawthorne wrote, "His inward trouble drove him to practice more in accordance with the old, corrupted faith of Rome, than with the better light of the church in which he had been born and bred. In Mr. Dimmesdale's secret closet, under lock and key, there was a bloody scourge." But when the adaptation shows us Dimmesdale reaching into his secret chest for the instrument of self-flagellation, it also depicts an icon of Christ on the cross that the Puritan minister has hidden away—a serious misreading, to my mind, of the above passage and of the Puritan attitude, even in travail, toward iconography.

Literalism extends to the script, written by Allan Knee and Alvin Sapinsley, which is almost pure Hawthorne, except for some added lines about "the wilderness" that seem to pay homage to the scholarship of Perry Miller. But, as the above example indicates, literalism can easily be carried to excess.

Hawthorne speaks of Dimmesdale walking about "perhaps actually under the influence of a species of somnambulism." Hauser seems to have taken this as a keynote for the direction of nearly all his players, who tend to behave like the hypnotized actors in Werner Herzog's film *Heart of Glass*. Not only is John Heard, as Dimmesdale, in an apparent daze, but Meg Foster as Hester Prynne is so low-key that she is often inaudible—a condition not helped by a tendency to insert music over some of her soft-spoken lines.

The only actor to escape suspended animation is Kevin Conway, who plays Roger Chillingsworth, the heartless man of science and reason. It may be that the creators of this production were able to give Chillingsworth life because he is more familiar a figure to modern minds. The adaptation seems never to have come to grips with Puritans as human beings. Perhaps the creators of this work, misapplying the emphasis of recent scholarship, believed that the Puritans had so exhausted their spirits from overattention to theology that they had all fallen victim to neurasthenia.

The larger problem of "The Scarlet Letter" is its own debilitating sense of self-importance. The National Endowment for the Humanities, the Corporation for Public Broadcasting, Exxon Corporation, the Andrew W. Mellon Foundation, and the Arthur Vining Davis Foundations together put more than $2 million into the project. No doubt everyone was anxious that it be "quality" television, that it be historically accurate and true to Hawthorne, that it advance the appreciation of American literature, and so on.

Too many eggs in one basket. Who knows how soon those sources will again cough up such money for another literary adaptation? "The Scarlet Letter" may be a classic example of a situation where the negative purpose of determining not to fail guarantees there will be no success.

Friendly Fire

For a few weeks three or four years ago, whenever the latest *New Yorker* magazine arrived, I felt compelled to put everything else aside and read the serialized installment of C. D. B. Bryan's riveting nonfiction work, "Friendly Fire"—the account of an Iowa farm couple's efforts to discover how its elder son had been killed in Vietnam. Subsequently published as a book, Bryan's work has now been turned into a three-hour, made-for-television movie on ABC stations.

"Friendly Fire" is network television's first noteworthy contribution to the recent spurt of activity in the popular media aiming to re-experience and re-interpret the American involvement in Vietnam.

What was remarkable about Bryan's work, amid the welter of books and articles about Vietnam at the time, was the glimpse it provided into the American heartland, far from the loci of power and protest, and into the souls of people who were moved by personal tragedy to suffer the further pain of alienation from friends, neighbors, and the United States government. Bryan was also honest enough to depict his own confused and ambivalent involvement with his subjects, and their additional alienation from him.

The producers of "Friendly Fire" have chosen to focus their adaptation on the drama of personality. The script, by Fay Kanin, and the direction, by David Greene, hardly ever widens its horizons beyond the personal struggles of Peg and Gene Mullen, whose first son was killed in Vietnam. In the role of Gene, the production company (Marble Arch Productions) cast the versatile supporting actor Ned Beatty. The role of Peg went to Carol Burnett, in her first major television appearance since the demise of her variety series.

"Friendly Fire" is only ambiguously a star vehicle for Burnett. Though one of the great comic performers of our time, she has not had notable success (in such films as *Pete 'n' Tillie* and *The Front Page*) as a dramatic actress. Until, perhaps, now.

What Burnett brings to the part of Peg Mullen is a certain lightness of spirit, an easy familiarity, an open vulnerability that bind us to her character as she becomes increasingly single-minded, implacable, even bitterly committed to her viewpoint and her goal.

Much the same description could be used for Ned Beatty's performance. In many of his supporting roles, this fine actor's screen persona has been so self-effacing that it is possible to overlook his contribution. In a co-starring role, however, his low-key strengths are evident.

Who are Peg and Gene Mullen? They are farm folk from La Porte City, Iowa (population just over 2,000), who one late summer day in 1969 saw their son, Michael, fly off to Vietnam. Almost six months later to the day, they awaited the return of his body. Michael Mullen was one of more than 50,000 American men who were to die in Vietnam. So what reason did the Mullens have to express their grief and anger differently from thousands of other bereaved parents?

Their behavior after Michael's death cannot properly be "explained," and the production, to its credit, does not try. What "Friendly Fire" might have done, however, is present in greater depth the nature of the military attitude toward the soldier's death—an attitude that precipitated the Mullens' vehement response. The production at least might have expressed more clearly than it does that Michael's death, like much else about Vietnam, seems to have been regarded by the military primarily as a problem in public relations.

Michael Mullen was not counted among the American dead the week he was killed because his death was classified as "non-battle" (nearly one out of five American deaths was "non-battle"). His parents were told only that he was killed by "friendly fire."

In the television production, as was the case in real life, the Mullens' efforts to find out more details are constantly met with bureaucratic evasion, insensitive errors, and an aura of secrecy that soon convince the couple that the true circumstances are being covered up. While pursuing their seach for the facts about Michael's death, the Mullens, almost against their will, become anti-war activists. Their conservative neighbors do not approve of them; their surviving children resent their parents' seemingly morbid attachment to their dead sibling.

All of this feels true to the life of those years—a time when many people were impelled to question firmly held beliefs, or to undertake unprecedented acts, not so much from conviction or rational choice but because of the blunders of authorities and the incomprehension of peers.

The Mullens' unprecedented act is to purchase a full-page advertisement in a Des Moines newspaper on a Memorial Day and fill it with 714 crosses, representing the 714 Iowa men who, up to that point, had died in the war. This striking gesture out of the homeland of Nixon's "silent majority" brings to the Mullens national media attention and also the solicitations of C. D. B. Bryan, a novelist who had once taught at the University of Iowa Writers' Workshop.

Bryan's character makes his first appearance rather late in the television program—we are not previously told that the Mullens' story is his narrative and interpretation—and he appears as something of a self-appointed deus ex machina. He determines to find out what the Mullens cannot: what the United States government will not tell them—how Michael Mullen really died.

Bryan is portrayed by another fine actor, Sam Waterston, who played Nick Carraway in *The Great Gatsby* and a controversial Hamlet at New York's Lincoln Center. Waterston's screen persona seems natural for the part of the observer, the passive man who can do no more than understand. Just how far the Mullens have progressed in their obsession is made clear by the contrast between their attitude and that of their new friend and champion.

Bryan seeks to ease the Mullens' pain, to get the facts "so they'll be able to put it to rest, once and for all." He doesn't realize that the Mullens' feelings have moved far beyond the point where one fact or another would make a whit of difference; the government's actions in response to them have already created a new set of "facts" that can never be put to rest. Neither does he anticipate that his own penetration of the mystery of Michael's death will saddle him with a complicity of knowledge.

Bryan discovers how the accident that caused Michael's death came about. He sees it in all its many-sided pathos; he describes it as "just a stupid, senseless act of war that happens occasionally." He becomes almost godlike in his ability to understand the flaws of men at war and in the bureaucracies, to recognize their pains and troubles, and to forgive them.

The abyss between Bryan and the Mullens opens. The Mullens do not want his transcendent vision. They reject the hubris of his philanthropy. "I guess I don't know what you want from me," he complains, unwilling to grant that they want total commitment from him or they want nothing at all. Having presented them his gift of fact, Bryan is confused by their rejection. "The truth is just the truth," he says. "I don't want to have to choose sides." In response to this Peg Mullen can only say, "When you lose your son there's only one side."

At the end, the television production has one serious flaw. Ultimately it seeks to narrow and reduce the Mullens, to leave us with an image of them only as distraught parents (perhaps a little more distraught than normal, but justifiably so). We are meant to see them as emblematic. They seem to hold on to their anger—even against Bryan's presumed greater wisdom—in order to keep the flame of memory alive, not just for Michael, but for all who gave their lives in Vietnam.

This resolution is empty rhetoric. In Bryan's original work, the Mullens' anger grew to encompass far more than the fact of Michael's death. It grew to take in a government, and the way that government dealt with its citizens. The producers of the television adaptation have let us see this, but they have flinched from clarity and particularity. They have chosen in the end to universalize the Mullens, and thereby to neutralize them.

¿Que Pasa, U.S.A.?

If public television didn't exist, it would behoove the commercial networks to invent it. That is a fair characterization of the attitude that the networks hold about public TV. Since the public system presents such high-toned events as opera, dance, and drama, the networks need not bear the burden of producing unprofitable "quality" programs.

Meanwhile, the networks are pleased that the inveterate localism of public television—and the Congressional pressures that reinforce it—cripples the nonprofit system as a potential competitor. Since relativity few public stations broadcast the same program at the same time, the impact of publicity is dissipated and national ratings are reduced. Nevertheless, commercial broadcasters are annoyed that public television draws away from them a predominantly elite audience ("upscale" in television jargon) that might be particularly attractive to advertisers.

Public television's stance toward the networks is less easy to summarize. There is clearly some desire on the part of PBS to become a fourth network and to challenge the commercial broadcasters both in programming and in the ratings. Some Hollywood production companies have been invited to create programs for public television. Such prominent public-television productions as "The Best of Families" and "The Scarlet Letter" have been turned over for supervision to people temporarily borrowed from Hollywood or the networks (though the results were conspicuously unsuccessful). It's likely that public-television programming will grow to look more like the commercial product.

These reflections are engendered by the appearance on public-television stations of "¿Que Pasa, U.S.A.?," a 13-episode, half-hour situation-comedy series. A situation comedy on public television? Well, to be sure, a situation comedy with a difference. "¿Que Pasa, U.S.A.?" is billed as the first bilingual situation comedy on television. More precisely, perhaps, it's a Spanish-language program with enough English-language dialogue to explain what the jokes in Spanish are all about.

"¿Que Pasa?" is a production of public-television station WPBT-Miami and Community Action and Research, Inc., with financing from the U.S. Office of Education. I don't know the amount of the federal grant for the series, but it appears to have been generous, for the program is as professional in its design and production as anything that comes out of Hollywood.

With such a grant, the purpose of "¿Que Pasa?" is obviously

not mere entertainment. The series centers on the Peñas, a Cuban exile family living in Miami and experiencing the age-old problems of immigrants—how to adapt to the mores of the new country, how to preserve the valued customs of the old. Under one roof live three generations: the grandparents, Adela and Antonio; the middle generation, Juana and Pepe; and their two children, Carmen and Joe.

"¿Que Pasa?" may be expected to appeal primarily to Spanish-speaking viewers. That certainly is the impression created by the response of the members of the live studio audience, who laughed almost exclusively at the Spanish dialogue; the English dialogue, although occasionally witty, with jokes and word play familiar from network sit-coms, is intended more as exposition. For non-Spanish speakers, the show serves the function of conveying information, in a comedy format, about Hispanic peoples adapting to life in the United States.

Watching several episodes of "¿Que Pasa?" from both the beginning and the end of the series, I had the uncanny feeling of seeing something both familiar and strange. The creators of the program—José R. Bahamonde is project director, Bernard Lechowick is producer and director, Luis Santiero is story editor and principal scriptwriter—have obviously chosen to emulate in large degree the conventions of the network sit-com, from the opening theme song to the live audience to the half-hour-vignette story structure. And it cannot be a coincidence that actress Ana Margarita Meńendez, who plays daughter Carmen, is a dark-haired look-alike for Sally Struthers—Gloria in "All in the Family." Yet in its content the program moves subtly beyond the boundaries of the current Hollywood formulas.

In the first episode of the new series, an elderly woman visiting the Peñas suddenly feels ill, lies down on the sofa, and appears to fall asleep. Several moments later a family member discovers that the woman has died. The remainder of the show is concerned with the effects of the death on the Peña household and the larger Cuban community.

"Dying in this country is very disagreeable," says one Cuban. "People here die so alone."

Another viewpoint is expressed by a blonde Anglo schoolmate

of Carmen's, Sharon. "With all the friends and relatives you people have," she says, "you're going to spend your whole life in mourning."

The Peñas attend the funeral, where the deceased lies in an open casket and the mourners find an occasion to greet old friends, make new ones, and conduct some business. A PBS description of this episode says that the old woman "drops dead," language that suggests the usual sit-com buffoonery, but it isn't like that at all. Death is treated with a tact and a naturalness that are not at odds with the premise of the show.

A later episode concerns the varied religious beliefs of Cubans. The Peñas are having a run of bad luck, and someone suggests they are victims of witchcraft. Carmen, Sharon, and another Cuban schoolmate—all dressed identically in their parochial-school outfits—discuss the matter, and one of the Cuban girls says, "We don't believe in those things. We're Catholic." But the episode depicts the survival of Santería, the Afro-Cuban religion that combines aspects of Roman Catholicism with magical practices, among the Cuban émigré community.

Here again, although elements of sensationalism are present and occasionally exploited, what is striking about the episode is the natural way it admits religion into the daily lives of the Peña family. Such subjects as death and religion are not totally absent from the network sit-coms, but rarely are they treated as part of the normal life experience of the situation comedy characters.

Because of its unique nature, as a polished and entertaining comedy series that resembles the commercial product but moves beyond it in both obvious and subtle ways, "¿Que Pasa, U.S.A.?" raises questions about the situation comedy as a form of cultural expression.

Since Norman Lear began putting explicit social and cultural lessons into his situation comedies of the early 1970s, no one can be unaware that an apparently mindless form of entertainment is capable of conveying powerful ideological messages—overtly, as in Lear's case, but with equal strength in other, more implicit shows. This should hardly be startling news; as far back as World War I cultural commentators were recommending the "comedy-drama" motion picture (an earlier version of sit-coms) as the ideal vehicle for communicating "approved" social values in a palatable manner.

"¿Que Pasa?" falls very much within this framework. Its theme song announces its purpose: "Say Hello, America / We are part of the new U.S.A. / People, listen, people / Let us share what we have today." The title of the show and the lyrics of the song present the rhetoric of assimilation. The Miami Cubans—as the differences among the three Peña generations graphically show—are in transition from an old life to a new. Most of the episodes deal directly with the Cuban encounter with American ways.

I don't mean to depict this portrayal as sinister; I do mean to insist on its ideological nature. This is a series about the humorous trials and tribulations of an émigré family coping with the process of becoming "American."

Millions of us—and millions more of our ancestors—have undergone such an experience, but are the Peñas like the rest of us? Why did they leave Cuba? Are they in touch with family and friends who remained? Are they permanently Americans, or do they dream of going back? The programs provide no clues.

The series is bereft of politics and economics. The Peñas seem to be a working-class family—Pepe carries a lunch pail to work—but their concerns are exclusively cultural, centering only on conflicting types of social values and behavior.

The Office of Education has chosen to finance a program that depicts an important, but only a very small, part of the experience of Hispanic people in the United States. One wonders if other Hispanic groups, such as the Puerto Ricans of East Harlem, the Chicanos of the East Los Angeles barrio, or the emerging Mexican majority in San Antonio, Texas, find their own experience reflected in the situations the series presents.

"¿Que Pasa, U.S.A.?" is important for its bilingual format and for its admirable qualities as entertainment. But a situation comedy is a situation comedy, and its messages cannot be ignored. Even if you can't understand the Spanish, it's still possible to read between the lines.

Mary Tyler Moore

Once upon a time she was Laura Petrie, the wife of a television writer, in "The Dick Van Dyke Show." Later, in her own situation-comedy series, she became Mary Richards, the thirty-ish career woman working in a Minneapolis television newsroom. Most recently she has been Mary McKinnon, a television variety-show star, in an hour-long backstage comedy series. Her fictional characters have grown more and more successful. Yet, in real life, Mary Tyler Moore, one of the most popular female performers in the history of American television, has experienced a perplexing season of television failures.

Moore's difficulties began last September with "Mary," an hour-long variety program that was her chosen format for her return to weekly television following her voluntary cancellation of "The Mary Tyler Moore Show," after seven years, in 1977. CBS scheduled "Mary" on Sundays at 8 p.m. following "60 Minutes," an apparently strong position on an evening the network has consistently dominated. The premiere show drew respectable ratings and respectful reviews that expressed admiration for her courage in trying to revive variety programming and for her reliance on a troupe of unknown repertory players rather than big-name guest stars. The reviews saw promise for the show's popular appeal.

Thereafter it was all downhill. Viewers who had tuned in to the first show, perhaps out of curiosity, apparently had seen enough. In the following weeks they turned elsewhere. "Mary" plummeted to near the bottom of the weekly prime-time ratings list. By early October, the show was pulled by "mutual agreement" between CBS and her production company, with the added news that she would soon be back on the air in a new format—with big-name guest stars.

They must have worked with feverish haste at MTM Enterprises (the production company built on the success of "The Mary Tyler Moore Show" and headed by Moore's husband, Grant Tinker). In little more than two months, with a new producer and an almost entirely new cast, "The Mary Tyler Moore Hour" went before the video cameras.

When the premiere episode aired in March, the reviews were no longer respectful, and the ratings were little better. It was swiftly announced that "The Mary Tyler Moore Hour" would not return in the fall. The actress and her production company would go back to the drawing boards to come up with a new situation-comedy vehicle for the 1980 fall season.

The simplest way to assess Mary Tyler Moore's year of travail would be to list the flaws of "Mary" and "The Mary Tyler Moore Hour," and they would fill most of this page. But that, I think, would miss the point. The more important questions have to do with the nature of television popularity, and with the strategies of an actress who wishes to grow and change.

By now it is fairly clear that there are significant differences between the ways audiences perceive performers in motion pictures and in television. The most successful movie stars have been actors whose private personalities project into the roles they are playing. They are real people first, fictional characters second. We go to see Brando or Jane Fonda or Woody Allen—and rarely remember the names of the characters they are playing.

With television performers it's different. Perhaps because of screen size or image definition, perhaps because of the weekly-series format, the character becomes more prominent than the actor. We seem to gain an intimacy with the fictional figures whose lives we follow week after week, and Archie or Fonzie or Rhoda become more real to us, and more important, than the performers who play them.

Some television stars try to overcome this phenomenon by playing "themselves," or by having their names in the show's title, or by giving their fictional characters their own first names—as did Mary Tyler Moore with Mary Richards and Mary McKinnon. That, however, does not prevent them from developing a fictional personality that seems to take over and, in the minds of viewers, becomes their "real" one.

In nearly twenty-five years of weekly television Lucille Ball was essentially the same "Lucy." James Garner brings his "Maverick" persona to "The Rockford Files" and to his recent series of commercials for Polaroid cameras. There may be limitations of talent or motives of personal predilection in the adoption of such long-lived fictional characterizations, but there is also an element of compulsion. Changing roles is much more difficult in television than in movies.

To her credit, Mary Tyler Moore has not been content to remain in roles that brought her enormous television success. She may have been prescient enough to know that her character of the sweet subservient helpmeet to Dick Van Dyke would fast become an anachronism after that show left the air in 1966. It's more likely, however, that she recognized the role's limitations on her talent and wanted to grow beyond it.

Significantly, to attempt a change, Moore left series television for the next four years, appearing in movies and on Broadway—but without notable success. Nevertheless, those years were enough to make possible her change in role from dutiful wife to single woman. Since the original premise of her successful sit-com series was that her character was recovering from a thwarted love affair, audiences may have fantasized that the fictional Mary was a divorced or widowed Laura Petrie when they heard the lyrics of "The Mary Tyler Moore Show" theme song: "How will you make it on your own?" and "You might just make it after all."

"The Mary Show," as it came fondly to be known, was a remarkable television phenomenon. Though it developed its audience slowly (Grant Tinker has often said that, under current network practices, the show would have been canceled in its first year), it eventually grew to command a rare kind of emotional attachment from its viewers—not the adulation accorded rock stars, but a quiet sort of loyalty and close feelings.

Mary's surrogate family of newsroom coworkers became, through years of familiarity, extensions of the viewer's own circle of friends. Perhaps no other situation comedy in television history has assembled so large and distinctive a group of supporting performers and fictional characters. Valerie Harper, Cloris Leachman, and Ed Asner carried on their roles in shows of their own—respectively, "Rhoda," "Phyllis," and "Lou Grant."

What this suggests is an unusual modesty and generosity on the part of the program's star, Mary Tyler Moore. Apparently she felt secure enough to share the limelight with her supporting cast. Looking at episodes of the old "Mary Show" (which is now broadcast during daytime hours in many cities), it becomes obvious that Moore did not dominate it: She was its moral center, the conductor through which its action flowed, but she often relegated herself to a subordinate position in the dramatic action.

Perhaps this restraint, no matter how beneficial to the format and continuing popularity of the show, played as large a part in the decision to end "The Mary Tyler Moore Show" as any other motive. Let's face it, nearly everyone else on the show seemed to have more "personality" than Mary—more edge, more bite, more abrasive life. The sweet, reticent, upright Mary Richards may eventually have seemed as bland and restricting a characterization as Laura Petrie.

But how to break the hold of Mary Richards on audience perceptions? This time there was to be no four-year hiatus from series television, no exploration of different show-business opportunities other than one modestly successful foray into television dramatic acting—"First You Cry," a docudrama in which Moore portrayed news broadcaster Betty Rollin, who was afflicted with breast cancer.

In returning so quickly to weekly television, Mary Tyler Moore wanted to be seen as the immensely successful and talented woman she was—to widen her range of dramatic emotions, to be a little sexy, to dance a lot (she had begun her career as a dancer). In "Mary," she wore her hair short and presented herself as a mature, confident woman.

The viewing public was not ready, however, to accept their Laura Petrie/Mary Richards looking like Shirley MacLaine and hoofing like Cyd Charisse. Nor, apparently, was it particularly attracted by the variety-show format. These problems were exacerbated by mediocre scripting by no fewer than eleven credited writers—seven of whom were retained among the even dozen who have done no better for "The Mary Tyler Moore Hour."

Grant Tinker's retrospective nightmare finally came true. CBS, which had been the comfortable ratings leader when "The Mary Tyler Moore Show" began, was running to stay out of third and last place when "Mary" faltered; there would be no grace period now. The failure of "Mary," however, sapped the courage from MTM Enterprises. The hasty reconstruction of the show into "The Mary Tyler Moore Hour" was based on the flimsy assumption that the public would accept a compromise version of the real Mary Tyler Moore—the Laura Petrie/Mary Richards fictional personality in the guise of a television variety-show star.

The format was a mistake from the beginning. It showed her behind-the-scenes preparation for "The Mary McKinnon Show," centering on the difficulties of finding a weekly guest star. It did give Moore a chance to dance at least one number a show and to sing and trade quips with her guests. But the inside show-business plot was tired from the start, and often grotesquely gauche, as when someone said to Dick Van Dyke (a guest on one program), "Don't you think Mary looks a lot like the gal that played Laura Petrie on your show?" and Van Dyke, shaking his head, said, "No."

That line (and others like it on the same program) may have been planted in a desperate effort finally to distance Mary Tyler Moore from her prior fictional characters. As the difficulties of the past season have shown, this is no easy task. Inadvertently, Moore's dilemma may be revealing the limitations of television acting. We will have to wait for her new situation comedy to know for sure.

Real People

Fred Silverman, as every sentient being in America must know by now, has been the dominant figure in television programming during the 1970s. As head of programming at CBS from 1970 to 1975, he strengthened that network's leadership in the ratings. Hired away by ABC, he presided over *that* network's astonishing leap to the top spot. In 1978 he was selected to be president of NBC—head not just of entertainment programming, but of the entire network—with the mandate to pull the operation out of its nosedive in popularity.

Early in 1979, halfway through his first season at NBC, Silverman was ready to junk nearly all of the prime-time schedule he inherited from his predecessors and to unveil a new slate of programs, hurriedly produced in Hollywood. Many of these—"Highcliffe Manor," "Whodunnit," "Sweepstakes," and "Turnabout," for instance—turned out to attract even fewer viewers than the shows they replaced. Like "Supertrain," another Silverman inspiration that proved an expensive fiasco, they seemed inept or cynical efforts to exploit borrowed formulas. Inevitably, Silverman's savvy was called into question, at least by the press.

Undaunted, Silverman replaced the replacements. It was clear that he was groping toward an approach to television programming that would capture the feelings of viewers entering the 1980s—just as such CBS programs as "Maude" and "Rhoda" and ABC shows like "Laverne & Shirley" seemed to express a mood of the times when Silverman put them on the air at the other networks.

Well, such a show finally seemed to emerge (at least as NBC tells it) in the spring. It's called "Real People," variously billed as a comedy show or a "humor magazine," and it's produced by George Schlatter, who was executive producer of "Rowan & Martin's Laugh-In," the iconoclastic variety show that was a big hit in the late 1960s. Six "Real People" programs were broadcast, and the show is scheduled to return in the fall as a regular series on the network.

Silverman gave special prominence to "Real People" when he announced NBC's fall schedule at the annual meeting of affiliated stations in Los Angeles. "As an innovative concept," he said, "it represents the direction in which NBC must move."

His remarks seemed to point toward a set of assumptions about what the television audience is now looking for: programs that are involved with the contemporary world, programs that do not take themselves too seriously. Putting this another way, NBC is trying to distill the essence of "Saturday Night Live" and make it work in prime time.

"Saturday Night Live" is a program for people who both love television and disdain it; it allows viewers to express both feelings at once. They can enjoy the show and also take pleasure in the fact that it parodies and puts down conventional television.

An intricate psychological process is at work here—a process that commercials (which are what television, after all, is all about) have already developed to a high degree. It involves a direct address to the audience, a recognition that the viewing experience is simultaneously distant and intimate.

In movies, audiences feel intimacy by identifying with the performers on the screen, but that seldom happens in television. Instead, television performers say to their viewers: We are separate, but we are in this together; we all recognize the foolishness of what we are doing, but we can agree to relax and let the artifice proceed. In this manner commercials break down the viewer's resistance to their sales pitch, and "Saturday Night Live" creates the impression that studio audiences and home viewers are somehow participating in the production of the show.

Perhaps because it is new and still unsure of itself, "Real People" makes these devices appear quite blatant—for example, by asking members of the studio audience to read the cue cards announcing commercial breaks. Or perhaps the lack of subtlety is intentional, thought to befit a program that reaches out to the prime-time mass audience.

The influence of "Saturday Night Live" is only one part, however, of the mix of styles that shape "Real People." The program is also one part daytime game show and one part local news.

The premise of the show is that it is fact, not fiction—entertainment and comedy created, or at least contributed, by the "real people" out there in America. "You're our stars," reads the ad for "Real People" in *TV Guide,* ". . . and we've tracked you down, all across these United States. 'Cause we're fascinated by the funny, free-wheeling antics of the folks next door."

Like Schlatter's "Laugh-In," "Real People" cuts swiftly from one segment to another; on the several shows I have seen, there were a dozen or more separate segments during the hour-long program. The majority of these were on film or videotape, but they are orchestrated by the studio portion of the show, which was broadcast live each week from the NBC studios in Burbank, California.

There are no fewer than five hosts. The most familiar is probably Fred Willard, who played opposite Martin Mull on "Fernwood 2-Night" and "America 2-Night," parodies of late-night talk shows that Norman Lear briefly produced as successors to "Mary Hartman, Mary Hartman." The other hosts are Sarah Purcell, John Barbour, Skip Stephenson, and Bill Rafferty.

The co-hosts begin the shows by fanning into the studio audience and inviting people to offer wisecracks about current events, such as the rapid rise in gasoline prices. I was skeptical about this, because on the shows I watched the members of the audience who were called upon seemed too handsome and glib to be "real people." I thought they were actors, but then maybe I have less faith in the comic genius of the folks back home than the producers do.

On another studio segment, one of the hosts showed examples of funny typographical errors, odd road signs, and the like, sent in by members of the viewing audience, who receive a "Real People" T-shirt for their endeavors. Both programs I saw included an off-color story about marriage, narrated in turn by all five of the hosts, and a brief bit by comedian Mark Russell from Washington. Journalist Jimmy Breslin also offered a commentary.

The film and video segments were generally capsule feature stories, shot on location with one of the hosts as "reporter." They included a 97-year-old editor of a weekly newspaper in Nevada who fulminates against gambling; a California man who likes to eat dirt; a school for beggars in New York City; and "Disco Harry," a 66-year-old in Ft. Wayne, Indiana, who dances the night away while his wife works a factory swing shift.

That is the "local news" aspect of "Real People"—you can see little stories such as those every night of the week on local television news shows anywhere in the country. It's daring of "Real People" to put these vignettes on prime time. The strongest element of the show may be its willingness to say that you people out there *are* interesting, and we want to show your faces on the home screen. Andy Warhol is reputed to have said that everyone will be a celebrity for fifteen minutes; "Real People" seems willing to give us at least four or five.

But if "Real People" has taken over from the local news the function of rescuing some of us from anonymity, it also perpetuates the least appealing of the local news formulas—the framing of the little features not to illuminate human variety, but to package it and put it down. Most of the characters who appear on "Real People" were made fun of, made to appear as butts of our superior wisdom—sometimes by the glaring device of cutting to shots of people laughing or scoffing or rolling their eyes.

"Real People" as yet has no consistent tone. Even the final program this spring was subject to so much last-minute tinkering that, of the five segments listed as "highlights" in publicity material, only two appeared on the air. If this is the wave of NBC's future, I hope it can grow beyond the shallow exploitation of American oddballs. "Saturday Night Live" at least pokes fun at itself and at television (including, on one memorable occasion, NBC's Fred Silverman) before it turns its wit on us.